VCR IC Data Files

VCR IC Data Files

John Edwards

Newnes

OXFORD BOSTON JOHANNESBURG MELBOURNE NEW DELHI SINGAPORE

Butterworth-Heinemann
Linacre House, Jordan Hill, Oxford OX2 8DP
225 Wildwood Avenue, Woburn, MA 01801-2041
A division of Reed Educational and Professional Publishing Ltd

A member of the Reed Elsevier plc group

First published 1998
Transferred to digital printing 2004

British Library Cataloguing in Publication Data
A catalogue record for this book is available from the British Library

ISBN 0 7506 3993 8

Library of Congress Cataloguing in Publication Data
A catalogue record for this book is available from the Library of Congress

Typeset by Avocet Typeset, Brill, Aylesbury, Bucks

PLANT A TREE

British Trust for Conservation Volunteers

FOR EVERY TITLE THAT WE PUBLISH, BUTTERWORTH-HEINEMANN WILL PAY FOR BTCV TO PLANT AND CARE FOR A TREE.

Preface

As domestic home entertainment equipment such as televisions, video recorders and hi-fis become more and more sophisticated but cheaper to purchase, the consumer electronics engineer must be armed with a high degree of technical knowledge and quick access to service information relating to the equipment under repair. It is the latter in which the independent service engineer is badly served by the industry. A fast, economic repair is the only way a repair business can remain viable. The majority of manufacturers' technical departments will not supply technical assistance to a non account engineer struggling with a difficult fault symptom. Service manuals are usually too expensive, no longer available or poor quality photocopies, and all too often the information they contain does nothing in the way of explaining how the circuits function.

All modern electronic equipment uses dedicated integrated circuits. In this book you will find the most common devices (and some not so common), encountered in the day-to-day repairs of video recorders. The signal routes in and out of the devices are also shown. Measurements were taken under actual working conditions. Where it's considered helpful, voltage readings, also taken under actual working conditions, are included. Some measurements will vary depending on the associated circuitry and test equipment used. It is also true that some devices can be used in various ways, but within these pages are the most common applications you will encounter and only minor variations will be found. A brief working description of the device is also given, so if you don't recognize a device just look it up.

This book is not intended to explain the intricate internal workings of the devices, that would in any case be of little use in practical fault finding, and as much as possible, plain language has replaced 'in house' jargon and confusing pin abbreviations so often used by i.c. and set manufacturers.

As a useful `memory refresher', and to give meaning to the 'pin function' description of the devices, I have included diagrams in block form showing the basic arrangements of the luminance and chroma sections, a typical capstan and drum motor drive circuit, servo, and tuner/IF. Also, a few commonly encountered, but rarely remembered, logic devices.

This book was written by a practising engineer of over 25 years' experience as an aid for the engineer of today. This and future editions of *VCR IC Data Files* and its companion *Television IC Data Files* will build up to a valuable reference source, taking the frustration out of not knowing what signal should be where during fault finding.

Abbreviations used

ACE head	Audio and control pulse head assembly
AFC	Automatic frequency control (normally associated with the IF section)
AFT	Automatic fine tuning (normally associated with the IF section)
AGC	Automatic gain control (normally associated with the Tuner/IF section
Carriage motor	The motor that raises and lowers the tape cradle when a tape is inserted or ejected
CR	Capacitor and resistor circuit usually representing a tuned or resonant circuit
E-E	The 'off air' or external input video and sound signal available when in the tape stop mode
Flip flop	A switching pulse, usually converted to a symmetrical square wave, obtained from the PG or FG pulse heads picking up a pulse from a magnet attached to part of a rotating motor
FM	Off tape playback or recorded signal. (Frequency modulated waveform)
H or L, Hi or Lo	This represents one of two states, either L (Low) or H (High). Normally H would indicate a voltage near the available supply rail and L would indicate a voltage near zero or much lower than the H state
kHz	Thousand cycles per second
LC	Coil (inductor) and capacitor circuit usually representing a tuned or resonant circuit
Loading motor	The motor that operates the deck mechanism i.e. tape lace up around the heads, shifting gears, turning mode switches, etc.
LP	Long play speed mode
MHz	Million cycles per second
Micro	Microprocessor
n/c	Not connected.
Pk-Pk	Peak to peak amplitude voltage of a waveform
SP	Standard speed play mode
Sync	Normally referring to Horizontal (Line) and/or Vertical (frame) synchronization pulses
Sys control	System control circuit. Usually a custom made micro processor
Trick speeds	Capstan and or drum motor operation during speeds other than normal SP, i.e. search, review, pause, jog/shuttle and still modes.

Video head 1 or 2	Video heads, sometimes referred to as channel A or B. Also as channel 1 or 2
4.43MHz xtal	Crystal oscillator, this one is 'ringing' or 'tuned' at 4.43MHz

For clarity, Logic gate devices as used for signal selection and system control switching, etc., are described as if they are their counterpart mechanical device.

Voltage measurements are indicated in the following form:

0v2 = 200mV (milli–volts), 0.2v
2v5 = 2.5V
5v0 = 5v

I prefer this method because it prevents confusion should the decimal point go missing during the printing process!

AJ14P

RF convertor

Package type A (9 pins)

Pins		Function
(1)	5.5MHz Oscillator LC circuit	
(2)	5.5MHz Oscillator LC circuit	
(3)	←	Audio input
(4)	→	Modulated RF output
(5)	0v Ground	
(6)		
(7)	←	Composite video input
(8)	→	Modulated RF video output
(9)	←	9v Supply

AJ7016

RF convertor

Package type B (12 pins)

Pins	Function
(1)	Audio modulator LC circuit
(2)	n/c
(3)	Audio modulator LC circuit
(4)	⟵ Audio input
(5)	⟶ Modulated audio output
(6)	0v Ground
(7)	⟶ Modulated video output
(8)	⟵ Video input
(9)	⟵ Test signal on / off control
(10)	⟵ 9v Supply
(11)	Test signal oscillator LC circuit
(12)	Test signal oscillator LC circuit

AN3158

Dual operational amplifiers

Package type B (8 pins)

Pins		Function
(1)	⟶	Amplifier 1, output
(2)	⟵	Amplifier 1, inverted input
(3)	⟵	Amplifier 1, non inverted input
(4)	0v Ground	
(5)	⟵	Amplifier 2, non inverted input
(6)	⟵	Amplifier 2, inverted input
(7)	⟶	Amplifier 2, output
(8)	⟵	40v Supply voltage

The amplifier 1 or 2 designation is intended only to indicate the separate devices. This device is often used in RF circuits as a tuning voltage buffer. Hence the high supply voltage.

The other amplifier (normally amplifier 2) is not connected.

AN1458

Dual operational amplifiers

Package type B (8 pins)

Pins		Function
(1)	⟶	Amplifier 1, output
(2)	⟵	Amplifier 1, inverted input
(3)	⟵	Amplifier 1, non inverted input
(4)	0v Ground	
(5)	⟵	Amplifier 2, non inverted input
(6)	⟵	Amplifier 2, inverted input
(7)	⟶	Amplifier 2, output
(8)	⟵	12v Supply

Amplifiers designated numbers, i.e 1,2, are given to distinguish between the individual devices

Sound I.F amplifier & detector

Package type B (14 pins)

Pins		Function
(1) 2v0	←	Sound I.F
(2) 2v0	←	Sound I.F
(3) 0v Ground		
(4) 0v Ground		
(5)	←	12v Supply
(6)	←	Manual preset volume control
(7)	←	Mute
(8) 8v5	→	Audio output
(9) 3v8		
(10) 3v8		
(11) n/c		
(12)	→	Amplifier output
(13) n/c		
(14)	←	Amplifier input

Pins 12 & 14 are connected to an independent voltage amplifier. The input (pin 14) may be connected to pin 12 for further amplification or used as an entirely independent amplifier. Likewise pins 12 & 14 may not be used.

AN3130

RF convertor

Package type A (14 pins)

Pins	Function
(1)	FM modulator LC circuit
(2) ←	Audio input
(3)	FM modulator LC circuit
(4) ←	12v Supply
(5)	
(6)	
(7)	
(8)	0v Ground
(9)	ac coupled to pin 11
(10) ←	Composite video input
(11)	ac coupled to pin 9
(12)	
(14) →	RF output

AN3171S

Colour decoder

Package type B (28 pins)

Pins		Function
(1) 0v Ground		
(2)	←	1v0 Chroma
(3) 2v5		
(4) 2v5	→	Chroma
(5) 2v8	←	Tint control
(6) 3v1		
(7) 2v8		
(8)	←	Phase control
(9)		
(10) 3v3	←	Red - luminance
(11) 3v3	←	Blue - Luminance
(12)		
(13) 4v2		
(14)	←	5v Supply
(15) 3v3	←	Video
(16) 0v4	→	Composite sync' pulses
(17)	→	Demodulated red - Luminance
(18)		
(19) 2v8	→	Demodulated blue - Luminance
(20) 2v6		
(21)	→	PAL or NTSC identification pulse
(22)	←	Blanking pulse
(23) 2v7	→	180 degree carrier
(24)		
(25) 2v7	→	90 / 270 degree carrier
(26)	←	PAL = H, NTSC = L
(27) 2v2, xtal oscillator		
(28)	←	Colour level control or grounded

AN3181S

NTSC & PAL encoder

Package type B
(24 pins)

Pins			Function
(1)	0v Ground		
(2)	2v4		
(3)	2v5	→	Chroma burst signal
(4)	3v0		
(5)		←	Clamp & Blanking pulse
(6)	4v2	←	Composite sync'
(7)	5v0		
(8)	5v0	←	Delayed luminance
(9)			
(10)		←	Luminance
(11)			
(12)		←	5v Supply
(13)	1v7	←	Chroma burst signal
(14)			
(15)	2v5	→	Chroma burst signal
(16)	2v0	→	Video
(17)			
(18)	1v8	←	Luminance & chroma
(19)	1v8		
(20)	1v8	←	Blue - Luminance
(21)	1v8	←	Red - Luminance
(22)	1v8		
(23)	1v8	←	180 degree sub carrier
(24)	1v8	←	90 / 270 degree sub carrier

Luminance Playback & record processor

Package type B (28 pins)

Pins		Function
(1) 2v8	←	Horizontal sync' pulse from pin 2
(2) 0v4	→	5v pk-pk Horizontal pulse to pin 2 & chroma 5.06MHz circuit
(3) 2v5		
(4) 2v7	←	Record mode composite video input
(5) 2v2		
(6)	←	Playback video input
(7) 1v5		
(8) 1v7	→	Record & playback Luminance to pin 15 via Low pass filter
(9) 0v Ground		
(10) 2v0	→	Record & playback, 2v pk-pk Composite video
(11)	→	Playback video mute & artificial vertical sync'
(12)	←	Playback video & E-E / PB control switch, E-E = H
(13)		
(14)		
(15)	←	From pin 8
(16) 2v7	→	Record & playback luminance to pin 19 via picture detail enhancement circuits
(17)		
(18)		
(19) 4v0	←	From pin 16
(20)	←	5v Supply
(21)	→	Record mode Luminance to pin 23 via emphasis circuits
(22) 2v1		
(23)	←	From pin 21
(24) 2v7		
(25) FM modulator circuit		
(26) FM modulator circuit		
(27) 2v7	←	White & dark clip circuit
(28) 3v2	→	Record mode FM luminance out

AN3231

Luminance processor

Package type B (48 pins)

Pins		Function
(1) 3v0		
(2) 0v5	→	4v pk-pk composite sync' pulses
(3) 3v5, E-E gain preset		
(4) 2v6	←	1v pk-pk record / playback composite video
(5) 2v0 AGC, detector decoupling		
(6) 1v9	→	Video to pin 8 via low pass filter
(7)	←	5v Supply
(8) 2v6	←	400mV pk-pk Luminance
(9) 2v6	←	400mV video delay
(10) 2v2, ac coupled to pin 11		
(11) 3v0, ac coupled to pin 10		
(12) 3v2	→	200mV pk-pk to luminance delay line
(13) Correlation		
(14) 2v3	→	800mV pk-pk record / playback Luminance video
(15) 0v Ground		
(16) 1v7	←	Luminance from pin 14
(17) 2v5		
(18) 2v1		
(19) 0v5		
(20)	→	600mV pk-pk record mode, to deviation control
(21) 2v0	←	From pin 20
(22) 2v7, Dark clip preset		
(23) 2v7, White clip preset		
(24) 2v8	→	Record mode 800mV pk-pk FM luminance
(25) 3v8	←	200mV pk-pk FM playback luminance from head amplifier
(26) 3v2		
(27) 4v8		
(28) 4v0		
(29) 4v0		
(30) 2v6,(31) 2v6		
(32) 1v4, Playback level control		
(33) 0v9	→	400mV pk-pk Demodulated playback luminance
(34)	←	Playback 5v supply
(35) 2v9	←	From pin 33
(36) 2v5		
(37) 1v5		
(38) 3v5		

(Continued opposite)

(39) 0v6
(40) 2v8
(41) 2v3
(42) 2v8 ← user picture control
(43) 2v0 ← 400mV pk-pk playback chroma
(44) 2v8
(45) 1v9 → 1v5 pk-pk Composite Video
(46) 2v4 ← Clamp input from pin 45
(47) 0v1
(48) 2v3 → Composite video to TV modulator & scart etc

AN3311

Video head playback amplifier

Package type B (22 pins)

Pins		Function
(1) 2v4	←	5v pk-pk, 25Hz head switching flip flop pulse
(2) 0v9		
(3)	←	FM from Playback head 1
(4)	←	FM from Playback head 2
(5)		
(6) 0v Ground		
(7)		
(8)	←	FM from Playback head 3, still & trick speed modes
(9)	←	FM from Playback head 4, still & trick speed modes
(10)		
(11)	←	Head normal or trick amplifier switch control
(12) 2v5		
(13) 2v5		
(14) 4v5	→	FM envelope detection
(15) 2v5	→	200mV pk-pk, 627kHz playback Chroma
(16)	←	Playback 5v Supply
(17) 4v7		
(18) 4v6		
(19) 0v Ground		
(20) 1v5	→	150mV pk-pk FM Playback Luminance
(21) 3v1		
(22) 3v0		

Luminance playback FM demodulation & Drop out compensation

Package type B (32 pins)

Pins		Function
(1)		
(2) 5v0		
(3)		
(4)		
(5)		
(6)		
(7)	←	5v Supply
(8)	←	Playback Demodulated Luminance from pin 9 via Low pass filter
(9)	→	400mV pk-pk Playback Demodulated Luminance to pin 8
(10) 0v Ground		
(11)		
(12)		
(13)		
(14)		
(15)		
(16)	←	100mV pk-pk Playback FM from pin 20
(17)		
(18)		
(19)	←	300mV pk-pk Playback FM from head amplifier
(20)	→	250mV pk-pk Playback FM to DOC delay line & pin 16
(21)	←	50mV pk-pk Playback FM from pin 20 via delay line
(22) 0v Ground		
(23) n/c		
(24)		
(25) 200mV pk-pk Playback Luminance noise component		
(26)	←	5v Supply
(27)	→	200mV pk-pk Luminance out
(28)	←	Picture sharpness control
(29)		
(30)		
(31)		
(32) n/c		

AN3331

Record & playback head amplifier

Package type B (20 pins)

Pins		**Function**
(1) 3v5	←	25Hz flip flop head switching pulse
(2) 5v5		
(3)	→	Record FM to heads
	←	Playback FM from heads
(4) 0v Ground		
(5)	→	Record FM to heads
	←	Playback FM from heads
(6)		
(7)	←	Playback mode 5v Supply
(8)		
(9)	→	Playback 627kHz chroma output
(10) 0v Ground		
(11)		
(12)	→	Playback FM luminance output
(13)		
(14)		
(15)		
(16) 0v Ground		
(17)	←	Record FM luminance input
(18)	←	Record mode 12v Supply
(19)	→	Record mode 627kHz chroma
(20) 4v0		

AN3334K

4 × Video head amplifier

Package type B (30 pins)

Pins		**Function**
(1)	←	25Hz Head switching flip flop pulse
(2) 0v8		
(3)	←	Playback FM from SP head 1
(4) 0v Ground		
(5)	←	Playback FM from SP head 2
(6) 0v8		
(7)	←	Playback mode 5v Supply
(8) 0v8		
(9)	←	Playback FM from LP head 2
(10) 0v Ground		
(11)	←	Playback FM from LP head 1
(12) 0v8		
(13) 0v Ground		
(14)	→	Record FM to SP heads
(15)		
(16)	→	Record FM to LP heads
(17)	←	Record mode 12v Supply
(18) 0v1		
(19)	←	Playback = L, Record = H
(20) 2v2	→	627kHz playback chroma output
(21)	←	Record mode 627kHz chroma input
(22) 0v Ground		
(23)	←	Record mode Luminance input
(24)	←	Head amplifier select
(25)	←	Playback mode 5v Supply
(26)	→	120mV pk-pk playback FM output
(27)	←	Portion from pin 26 for AGC purpose
(28)		
(29)		
(30)	→	FM envelope detection

AN3360

Video & HiFi head playback & record amplifier

Package type B (36 pins)

Pins			Function
(1)		←	25Hz Head switching flip flop pulse
(2)		←	SP or LP amplifier switching control
(3)	1v5		
(4)		←	FM playback from SP head 1
(5)		←	FM playback from SP head 2
(6)	2v4		
(7)	0v Ground		
(8)			
(9)		←	FM playback from LP head 1
(10)		←	FM playback from LP head 2
(11)	1v5		
(12)		→	FM record mode to LP heads
(13)		→	FM record mode to SP heads
(14)		←	HiFi audio from head 3
(15)	0v Ground		
(16)		←	HiFi audio from head 4
(17)		→	Record mode to HiFi heads
(18)		←	Playback mode = H, Record mode = L
(19)			
(20)		←	5v Supply
(21)			
(22)	1v8	→	Playback HiFi sound output
(23)	2v2	←	Record mode HiFi sound input
(24)	0v Ground		
(25)			
(26)		←	Record mode 5v Supply
(27)		←	Record mode chroma & Luminance
(28)	3v2	→	FM playback envelope detection
(29)	2v4		
(30)		→	400mV pk-pk, 627kHz playback chroma output
(31)	0v Ground		
(32)		→	300mV pk-pk FM playback output
(33)			
(34)		←	LP mode = L
(35)		→	LP or SP playback envelope data
(36)		←	5v Supply playback mode

AN3380

Video head record & playback amplifier

Package type B (22 pins)

Pins		Function
(1) 0v7		
(2) 3v2		
(3)		
(4) 2v5	→	2v0 pk-pk Record FM to video heads
(5) 2v4	→	1v5 pk-pk Record FM to video heads
	←	Playback FM from video head 1
(6) 0v Ground		
(7) 2v4	→	1v5 pk-pk Record FM to video heads
	←	Playback FM from head 2
(8) 2v4		
(9)		
(10)		
(11)	←	5v Supply
(12) 3v0	←	Record FM
(13)	←	Mode control, Playback = H, Record & E-E = L
(14)	←	25Hz Head switching flip flop pulse
(15) 0v Ground		
(16)	←	Record start = L
(17) 2v7	→	250mV pk-pk Playback colour & FM luminance to pin 19
(18) 0v Ground		
(19) 3v5	←	180mV pk-pk Playback FM luminance from pin 17 via equalization circuit
(20)		
(21)	→	Playback FM
(22)	←	Composite sync' to internal record AGC circuit

AN3494

Luminance & chroma noise cancellation drop out compensation

Package type B (30 pins)

Pins		Function
(1) 2v0	←	Playback Chroma input
(2)	←	5v Supply
(3)	→	Playback Chroma to pin 5 via chroma cross talk cancelling circuit
(4) 0v Ground		
(5) 3v0	←	Chroma from pin 3
(6) 2v0		
(7)	→	Playback chroma to pin 9
(8) 3v0		
(9)	←	Chroma from pin 7
(10)	←	Playback chroma noise cancel on / off control
(11)	→	Playback chroma out
(12)	←	Edit mode = H
(13) 0v Ground		
(14)	→	Record mode luminance
(15)	→	Record & playback mode luminance
(16) 2v5	←	25Hz Head flip flop pulse
(17) 4v9		
(18) 1v5	←	Luminance level adjustment
(19) 3v0	←	Record & playback luminance from pin 20
(20) 1v5	→	To pin 19
(21) 3v0	→	Record & playback luminance to pin 22
(22)	←	From pin 21
(23) 3v0		
(24) 1v8		
(25) 0v3		
(26)	←	Playback tape drop out detection
(27) 2v4	→	Luminance to delay line
(28) 3v5	←	Luminance from delay line
(29)	←	Record 5v Supply
(30) 3v5	←	Playback & record luminance input

AN3495K

Luminance & chroma noise cancellation and tape drop out compensation

Package type B (30 pins)

Pins			Function
(1)	3v0	←	Playback Chroma input
(2)		←	5v Supply
(3)	1v5	→	Playback Chroma to pin 5 via delay line & chroma crosstalk cancelling circuit
(4)	0v Ground		
(5)	3v0	←	Chroma from pin 3
(6)	1v8		
(7)	3v2	→	Playback chroma to pin 9
(8)	3v0		
(9)	3v0	→	Chroma from pin 7
(10)	0v8	←	Playback chroma noise cancel on / off control
(11)	1v5	→	Playback chroma output
(12)		←	Edit mode = H
(13)	0v Ground		
(14)	1v6	→	Record mode luminance to external FM modulator
(15)		←	25Hz Head switching flip flop pulse
(16)	2v5	←	Luminance level adjustment
(17)	4v9		
(18)	0v Ground		
(19)	0v Ground		
(20)	1v5	←	From pin 21
(21)	3v0	→	Record & playback luminance to pin 20
(22)	2v2		
(23)	3v0		
(24)		←	LP mode = H, SP mode = L
(25)		←	5v Supply
(26)	1v5	→	Playback luminance to drop out delay line circuit
(27)	3v0	←	Luminance from delay line
(28)		←	Super VHS mode = H
(29)	3v5	←	Playback & record luminance input
(30)	3v5	←	Playback & record luminance input

AN3810K

Motor drive **Package type B (24 pins)**

Pins		Function
(1)	⟶	Stator coil 1 current control
(2)	⟶	Stator coil 2 current control
(3)		
(4)		
(5)		
(6)		
(7)		
(8)		
(9)	⟵	Stator coils current sensing for torque control
(10)		
(11)	⟵	Trick modes compensation
(12)		
(13)	⟵	Servo drive
(14)	⟵	Switching bias from hall effect device 1, stator coil selection
(15)	⟵	Switching bias from hall effect device 1, stator coil selection
(16)	⟵	Switching bias from hall effect device 2, stator coil selection
(17)	⟵	Switching bias from hall effect device 2, stator coil selection
(18)		
(19)	⟶	25Hz PG pulse & 150Hz FG pulse to servo circuit
(20)	⟵	Stator coils common return
(21)	⟵	Stator coils current sensing for torque control
(22)		
(23)	⟶	Stator coil 3 current control
(24)	Stator coils common return resistor	

Drum motor drive

Package type B (18 pins)

Pins		Function
(1) 9v3	⟶	Stator coil 2 drive
(2) 9v3	⟶	Stator coil 1 drive
(3) 0v5		
(4) 0v1	⟵	Output current sensing from pin 18
(5) 0v Ground		
(6) 2v5		
(7) 2v5	⟵	Drive from servo circuit
(8) 2v6	⟵	300mV pk-pk from –ve hall effect device
(9) 2v5	⟵	300mV pk-pk from +ve hall effect device
(10) 2v6		
(11)	⟵	5v4 Supply
(12) 3v9		
(13) 3v9		
(14) 3v9		
(15)	⟶	FG & PG pulses
(16)	⟵	10v Supply
(17) 9v4	⟶	Stator coil 3 drive
(18)	⟶	To pin 4

AN3814

Drum motor drive

Package type B (18 pins)

Pins		Function
(1)	→	Stator coil 2 drive
(2)	→	Stator coil 1 drive
(3)	←	Motor current sensing feedback from pin 18
(4)		
(5) 0v Ground		
(6)		
(7)	←	Drum motor torque control
(8)	←	+ve 360mV pk-pk from Hall effect device
(9)	←	–ve 360mV pk-pk from Hall effect device
(10)		
(11)	←	5v Supply
(12)		
(13)		
(14)		
(15)	→	PG & FG pulses to Servo
(16)	←	14v Supply
(17)	→	Stator coil 3 drive
(18)	→	To pin 3

AN3815K

Drum motor drive

Package type B (18 pins)

Pins		Function
(1) 9v5	→	Stator coil 2 drive
(2) 9v5	→	Stator coil 1 drive
(3) 0v1	←	Output current feedback & torque sensing from pin 18
(4) 0v6		
(5) 0v Ground		
(6) 2v5		
(7) 2v5	←	Drive control from servo
(8) 2v5	←	+ve 350mV pk-pk pulses from Hall effect device
(9) 2v5	←	–ve 350mV pk-pk pulses from Hall effect device
(10) 2v6		
(11)	←	5v Supply
(12) 3v9		
(13) 3v9		
(14) 3v9		
(15) 1v2	←	FG & PG pulses
(16)	←	9v Supply
(17) 9v4	→	Stator coil 3 drive
(18) 0v1	→	Torque control to pin 3

AN3822

Capstan motor drive

Package type B (24 pins)

Pins		Function
(1) 2v6	→	2v2 pk-pk drive to stator coil 1
(2) 0v1	→	Motor drive current feedback to pin 20
(3) 2v6	→	2v2 pk-pk drive to stator coil 2
(4) 0v		
(5) 2v2	←	–ve Hall effect device 3
(6) 2v2	←	+ve Hall effect device 3
(7) 2v2	←	–ve Hall effect device 2
(8) 2v2	←	+ve Hall effect device 2
(9) 2v2	←	–ve Hall effect device 1
(10) 2v2	←	+ve Hall effect device 1
(11)	←	Capstan on = L, Reverse = H
(12) 0v Ground		
(13)	←	5v Supply
(14) 3v2	→	Hall effect devices drive
(15) 2v8		
(16) 2v6	←	Drive from servo
(17)	←	Capstan speed increase control = H
(18) 0v6		
(19) 0v5	←	Noise shift pulse
(20) 0v2	←	From pin 2
(21) 0v Ground		
(22) 0v5		
(23) 2v6	→	2v2 pk-pk drive to stator coil 3
(24)	←	14v Supply

AN3990

Audio playback & record amplifier

Package type B (18 pins)

Pins			Function
(1)	4v3	←	400mV pk-pk from Tuner or E-E
(2)	4v3		
(3)	4v3		
(4)	4v3	←	Playback audio from pin 18
(5)	4v3	→	Record mode audio to pin 11
(6)	0v1		
(7)	4v4		
(8)	0v Ground		
(9)	4v5	→	Audio to RF convertor
(10)		←	9v Supply
(11)	4v3	←	Record mode audio from pin 5
(12)	4v3		
(13)	4v3	→	2v pk-pk record mode audio to head
(14)		←	Audio mute on = 2v5, off = 0v7
(15)		←	Record / playback control, Record = 2v5, Playback = 0v7
(16)	4v1		
(17)	4v2	←	Audio tape playback
(18)	4v3	→	Playback audio to pin 4

AN3991

Audio playback & record

Package type B (20 pins)

Pins		Function
(1) 5v0		
(2) 4v9	←	Playback audio from ACE head
(3) 4v9		
(4) 5v0	→	Amplified audio to pin 5 via playback level adjustment
(5) 5v0	←	Audio from pin 4
(6) 5v0	→	Record and playback audio to pin 7 & 11
(7) 5v0	←	Audio from pin 6
(8) 5v0		
(9) 5v3	→	2v pk-pk Record mode to audio head
(10)	←	10v Supply
(11) 5v0	←	Audio from pin 6
(12) 5v0	→	Audio to external line socket & RF modulator
(13) 0v Ground		
(14)	←	Record mode automatic gain control
(15)	←	Mute control on = H
(16)	←	Record / playback control. Playback = L, Record = H
(17) 5v0	←	Audio from Tuner or external line socket
(18)		
(19)		
(20)		

AN3994

Record & playback audio amplifier

Package type B (22 pins)

Pins			Function
(1)	0v Ground		
(2)	5v6	→	Playback and record audio
(3)	5v5	←	Playback audio from pin 5
(4)	0v		
(5)	5v7	→	Playback audio to pin 3
(6)	5v5		
(7)	5v5	←	Playback from head
(8)	0v		
(9)		→	Playback mode = H , Record mode = L
(10)		←	Mode control. Record = 0v6, Playback = 11v4
(11)		←	Mute = L
(12)	5v5		
(13)	0v6	←	SP mode = L
(14)	0v	←	Audio from pin 2 for internal AGC detection
(15)	0v		
(16)	5v5	←	Tuner, E-E & External line input
(17)	5v5		
(18)	0v		
(19)	5v5	→	Record mode audio to head
(20)		→	Record mode, bias oscillator circuit 11v supply
(21)		←	12v Supply
(22)	0v		

AN5033

Pulse width modulated to analogue tuning convertor

Package type B (18 pins)

Pins		Function
(1)	←	30v pk-pk reference tuning pulse from pin 18
(2)	→	Analogue tuning voltage to tuner
(3) 0v Ground		
(4)	←	30v Supply
(5)	←	Channel tuning manual adjustment
(6)	→	Error tuning voltage feedback to micro'
(7)	→	Error tuning voltage feedback to micro'
(8)	←	5v Supply
(9)		
(10)	→	Power down detect & store to memory
(11)		
(12)	←	Power on reset = L
(13)	→	Band switching to tuner
(14)	→	Band switching to tuner
(15)	→	Band switching to tuner
(16)	←	Tuning Band control, UHF = L, VHF = H,
(17)	←	Tuning Band control, UHF = H, VHF = H
(18)	→	To pin 1

Tuning voltage & band switching

Package type B (16 pins)

Pins		Function
(1) 0v Ground		
(2) 0v		
(3)	←	32v Supply
(4) 3v8	←	Pulse width modulated tuning voltage from micro'
(5) 0v Ground		
(6) 1v0	→	Pulse width modulated tuning voltage to pin 7 via Low pass filter
(7) 0v7	←	Analogue tuning voltage from low pass filter
(8) 0v7	→	0 -32v tuning voltage to Tuner
(9)	←	Tuning band control, UHF = H, VHF = L
(10)	←	Tuning band control, UHF= L, VHF = L
(11)	←	12v Supply
(12)	→	UHFTuning band control to Tuner
(13)	→	Tuning band control to Tuner
(14)	→	VHFTuning band control to Tuner
(15) 12v		
(16)	←	13v Supply (derived from same source as pin 3)

AN5111

Vision & sound I.F detector

Package type B (28 pins)

Pins		Function
(1) 4v5	←	I.F from SAW Filter
(2) 0v Ground		
(3) 4v5		
(4) n/c		
(5) 2v7	→	AGC control to tuner
(6) 4v4	←	AGC control
(7) 4v0	→	Amplified vision I.F to pin 9 via sound carrier trap
(8) 0v Ground		
(9) 5v1	←	From pin 7
(10) 8v6	→	Automatic frequency control (AFC) to tuner
(11) AFC, LC circuit		
(12) 6v2		
(13) 6v2		
(14) AFC, LC circuit		
(15) 0v	←	Audio mute = H
(16) 0v Ground		
(17) 4v5	→	Composite video to pin 18 via sound trap
(18) 4v5	←	From pin 17
(19)	→	2v0 pk-pk video output
(20)	←	12v Supply
(21) 5v6	→	6MHz Detected sound
(22) 0v Ground		
(23) 5v0		
(24) 0v4		
(25) 4v8		
(26) 4v6		
(27) 8v7		
(28) 4v5	←	I.F from SAW Filter

AN5215

Sound I.F detector

Package type A (7 pins)

Pins		**Function**
(1) 3v6	←	Sound I.F input
(2) 3v6		
(3)	←	12v Supply
(4) 0v Ground		
(5) FM detector LC circuit		
(6) FM detector LC circuit		
(7) 4v0	→	Audio output

AN6041

Dual balanced modulator

Package type A (9 pins)

Pins			Function
(1)	6v3	→	Modulated 1 output
(2)		←	Supply
(3)	2v3	←	4.43MHz signal 1
(4)		←	Playback chroma LF signal 1
(5)	0v Ground		
(6)		←	Playback chroma LF signal 2
(7)	2v3	←	4.43MHz signal 2
(8)			
(9)	6v3	→	Modulated 2 output

AN608P

Video amplifier

Package type A (4 pins)

Pins			**Function**
(1)	5v1	⟶	Video output
(2)		⟵	9v Supply
(3)	2v2	⟵	Video input
(4)	0v Ground		

AN6168

Digital Drum & Capstan servo

Package type B (28 pins)

Pins			Function
(1)	0v Ground		
(2)	2v6	⟶	5v pk-pk pulse width modulated Capstan speed control
(3)	2v4	⟶	5v pk-pk pulse width modulated Capstan phase control
(4)		⟶	Record & playback mode control pulses
(5)	4v8	⟵	Vertical sync' pulse input
(6)		⟵	Pulse width modulated playback tracking pulses for SP, LP review & normal speed
(7)		⟵	Record pause control
(8)	0v Ground		
(9)		⟵	Record 9v supply
(10)		⟵	Long play mode = H
(11)		⟵	5v Supply
(12)			
(13)	0v Ground		
(14)		⟵	5v Supply
(15)		⟵	4.43MHz clock pulse
(16)		⟵	ACE head control pulses
(17)		⟵	Capstan FG pulses
(18)		⟵	Capstan reverse brake = H
(19)		⟵	Capstan cue & review speed modes = H
(20)	0v		
(21)	0v		
(22)		⟵	Still & slow modes control
(23)		⟵	5v Supply
(24)	1v5	⟵	200 Hz (normal speed) Drum motor FG pulses
(25)		⟵	25 Hz, 5v pk-pk flip flop head PG pulse
(26)	0v		
(27)		⟶	5v pk-pk Pulse width modulated Drum motor phase control
(28)		⟶	5v pk-pk Pulse width modulated Drum motor speed control

Pins 20, 21 may be grounded

AN6209

Audio record & playback amplifier

Package type B (22 pins)

Pins		Function
(1) 5v3	←	Playback from audio head
(2) 5v3		
(3) 5v4	→	Amplified playback to pin 4
(4) 5v4	←	From pin 3
(5) 5v3		
(6)	→	Record or playback audio
(7) 5v2	←	Record audio from pin 6
(8) 5v5	←	Playback audio from pin 5
(9) 5v5	→	Record or playback audio output
(10) 0v Ground		
(11)	←	Automatic gain control
(12)	←	10v Supply
(13)	←	5v4 (1/2)Supply
(14) 5v4	→	Record mode to audio head
(15) 5v4	←	Record mode, LP = L, SP = H
(16) 5v4	←	Record audio from pin 6 via record current adjustment
(17)	←	Audio mute = 9v0
(18)	←	Mode control, Record = H, Playback = L
(19) 5v3	←	Audio from Tuner or E-E
(20) 5v2		
(21) 5v1		
(22) 5v1		

AN6306

Video playback FM modulator & record processor

Package type B (22 pins)

Pins		**Function**
(1) 3v5	←	Playback luminance
(2) 1v6		
(3) 1v7	←	Record mode E-E video
(4) 3v6	←	E-E level adjustment
(5) 0v1	←	Horizontal sync' pulse
(6) 0v1		
(7)	←	5v Supply
(8) 4v5, FM modulator circuit		
(9) 4v5, FM modulator circuit		
(10) 3v0	←	White clip adjustment
(11) 3v6	→	Record mode.800mV pk-pk modulated FM
(12) 2v2	←	Luminance from pin 14 via preset deviation & carrier controls
(13) 2v5	←	Dark clip control
(14) 2v3	→	Record mode. 600mV pk-pk Luminance to pin 12
(15) 2v3	←	Main signal pre-emphasis feedback
(16) 1v8	←	Record mode, main video from pin 18
(17) 2v7		
(18) 2v2	→	Record video to pin 16, playback video to external sync' separator
(19) 0v Ground		
(20) 1v8	←	Record & playback video from pin 22
(21)	←	E-E = 5v, Playback = 0v
(22) 3v0	→	Record & playback video to pin20

AN6307

Video head Chrome & luminance record amplifier

Package type A (9 pins)

Pins			Function
(1)	7v1	←	300mV pk-pk luminance
(2)		←	12v Record mode supply
(3)	5v7		
(4)	7v0	←	40mV pk-pk chroma
(5)	2v0		
(6)	5v7		
(7)		→	Switching voltage, record mode
(8)		→	3v pk-pk to Video heads
(9)		→	Switching voltage, record mode

AN6310

Luminance FM record processor **Package type B (24 pins)**

Pins		Function
(1) 4v0	←	Composite video
(2) 7v0,	→	E-E preset adjustment
(3) 6v5	←	E-E preset adjustment
(4) 3v6		
(5) 3v3		
(6)	→	Composite sync' pulses
(7) 0v Ground		
(8) 3v3		
(9) 0v	→	Modulated FM to record heads driver via high pass filter
(10) 0v	←	White clip preset adjustment
(11) 0v		
(12) 0v	←	Dark clip preset adjustment
(13) 0v, FM modulator LC circuit		
(14) 0v, FM modulator LC circuit		
(15) 0v		
(16) 0v8	←	Luminance from pin 18
(17) 0v1		
(18) 8v5	→	Luminance to pin 16
(19) 4v6	←	Luminance from pin 22
(20)	←	12v Supply
(21) 3v6		
(22)	→	Luminance to pin 19
(23) 12v		
(24) 8v0	→	Composite video to low pass filter

AN6320N

Video Head playback amplifier

Package type B (14 pins)

Pins		Function
(1)	←	12v Supply
(2) 0v7		
(3)	←	Head 1 FM input
(4) 0v Ground		
(5)	←	Head 2 FM input
(6) 0v7		
(7) 6v5		
(8) 3v0	←	Head 2 gain control
(9) 8v8	→	Head 2 to Balance preset
(10) 7v5	→	Head 1 & 2 combined FM output
(11) 2v	←	Balance preset
(12) 1v0	←	25Hz Drum head switching pulse
(13) 9v0	→	Head 1 to Balance preset
(14) 3v1	←	Head 1 gain control

AN6326

Video head amplifier

Package type B (18 pins)

Pins		Function
(1)	←	5v Supply
(2) 2v3	←	25Hz Head switching PG pulse
(3) 0v8		
(4) 0v7	←	Head 2 Playback FM
(5) 0v Ground		
(6) 0v7	←	Head 1 Playback FM
(7) 0v8		
(8) 3v2		
(9) 3v2		
(10) 2v6	→	200mv pk-pk playback chroma
(11) 3v2		
(12) 1v6	→	Video via chroma trap to pin 13
(13) 3v2	←	300mV pk-pk Luminance
(14) 0v Ground		
(15) 5v0		
(16) 3v1	→	Drop out compensation detection
(17) 3v3	→	250mV pk-pk Luminance FM
(18) 3v1	←	60mV pk-pk Delayed drop out compensation

Video head FM sound playback preamplifier

Pins		Function
(1)	←	5v Supply
(2) 2v5	←	25Hz Head switching PG pulse
(3)		
(4)	←	Right channel playback FM sound
(5) 0v Ground		
(6)	←	Left channel playback FM sound
(7)		
(8)		
(9) 3v3		
(10) 2v8	←	100mV pk-pk playback sound FM to pin 12
(11) 3v2	→	FM playback sound output
(12) 1v7	←	From pin 10
(13) n/c		
(14) 0v Ground		
(15) n/c		
(16) n/c		
(17) n/c		
(18) n/c		

Demodulator & sync' separator

Package type B (22 pins)

Pins		Function
(1) 0v4	⟶	900mV pk-pk horizontal sync' pulses
(2) 4v2		
(3)		
(4)		
(5) 4v0	⟵	Playback luminance
(6) 4v0		
(7)		
(8)		
(9)	⟵	E-E = 5v, Playback = 0v
(10) 4v1		
(11) 4v1		
(12) 1v0	⟶	Playback luminance
(13)	⟵	5v Supply
(14) 2v9	⟶	To noise limiter circuit
(15) 2v8	⟵	From noise limiter circuit
(16) 4v0	⟵	High frequency noise from playback luminance signal
(17) 2v9	⟵	4.43MHz playback chroma, pin goes low for mute (screen - blanking)
(18) 3v3	⟵	Sharpness control
(19) 2v4	⟶	Record & playback mode
(20) 3v0	⟵	Record (E-E) mode composite video
(21) 0v Ground		
(22) 4v3	⟵	Video to internal sync' separator circuit, output at pin 1

AN6337

Luminance demodulator

Package type B (22 pins)

Pins		Function
(1) 0v7	⟶	3v0 pk-pk Vertical sync' pulse
(2) 3v8		
(3) 1v9		
(4) 1v9		
(5) 3v7	⟵	Playback mode FM from drop out compensation circuit via High pass filter
(6) 3v6	⟵	Playback mode FM from drop out compensation circuit via low pass filter
(7) 2v4		
(8) 2v4		
(9)	⟵	Mode control, Record = L, Playback = H
(10) 3v2		
(11) 3v2		
(12) 2v0	⟶	Demodulated FM
(13)	⟵	5v Supply
(14) 2v0		
(15) 2v6		
(16) 3v5		
(17) 2v6	⟵	250mV pk-pk, 627kHz Chroma
(18) 2v0	⟵	1v0 pk-pk Playback video from demodulator via de-emphasis & picture sharpness control
(19) 2v3	⟶	1v5 pk-pk E-E from pin 20 & playback video output
(20) 3v0	⟵	300mV pk-pk Record mode video from Tuner & line input
(21) 0v Ground		
(22) 3v8	⟵	3v0 pk-pk Playback video to internal sync' separator. Output at pin 1

AN6341

Capstan motor servo control

Package type B (16 pins)

Pins		Function
(1)	←	12v Supply
(2) 0v8	←	SP = H, LP = L
(3) 1v5	←	Set normal speed preset
(4) 1v5		
(5) 0v5		
(6)		
(7) 2v8	→	Capstan motor phase control output
(8) 0v Ground		
(9) 5v0	→	Capstan motor speed control output
(10) 5v5		
(11) 3v3		
(12) 4v7		
(13) 5v6	←	12v pk-pk sample pulse
(14) 0v4	←	Tracking control
(15) 0v3	←	PG Pulse
(16) 7v9	←	1v pk-pk FG pulse

AN6346N

Video head switching pulse generator and record/playback ACE head control pulse amplifier

Package type B (18 pins)

Pins		Function
(1)	⟶	Amplified ACE head playback control pulse to Capstan servo
(2) 0v3	⟵	From pin 3
(3)	⟶	2v5 pk-pk ACE head playback control pulse to pin 2
(4)	⟵	Capstan reverse = H
(5) 2v7		
(6)	⟵	Control pulse from ACE head
(7) 0v Ground		
(8) 2v6		
(9)	⟵	Playback = L, Record = H
(10)	⟶	Record mode control pulse to ACE head
(11)	⟵	5v Supply
(12)	⟶	Pulse width modulated control pulses to capstan servo
(13)	⟵	Manual tracking control
(14)	⟶	5v pk-pk, 25Hz head switching pulse to Drum servo
(15)	⟵	Drum PG 1 manual adjustment
(16)	⟵	Drum PG 2 manual adjustment
(17)	⟵	300mV pk-pk Drum PG pulse
(18) 0v Ground		

AN6350

Drum motor servo control

Package type B (28 pins)

Pins			Function
(1)		←	12v Supply
(2)	7v5	→	Capstan playback (PG pulse) SP = 3v5 pk-pk, LP = 3v pk-pk
(3)	0v Ground		
(4)	1v5		
(5)	0v Ground		
(6)	1v5	←	ACE head control pulse
(7)	0v Ground		
(8)	6v0	→	Drum motor drive speed control
(9)	5v5		
(10)	0v2		
(11)	2v9		
(12)	0v Ground		
(13)	1v5		
(14)	3v4	←	Drum motor FG pulse
(15)	5v0	→	Drum motor drive phase control
(16)	5v6		
(17)	3v4		
(18)	5v5		
(19)	0v4	←	Record SP phase control
(20)	5v5	←	10v pk-pk head switching pulse
(21)	5v5	←	Head switching point control
(22)	4v9	←	Head switching point control
(23)	–1v0	←	+ve Drum motor PG pulse
(24)	12v6	←	–ve Drum motor PG pulse
(25)	7v4	←	7v pk-pk Vertical sync' pulse
(26)	7v5		
(27)	2v5		
(28)		←	Record = 12v, Playback = 0v

AN6356

ACE head playback/record control pulses & Video head flip flop generator

Package type B (18 pins)

Pins			Function
(1)		⟶	5v pk-pk amplified playback ACE head control pulse to servo
(2)		⟵	Amplified playback ACE head control pulse from pin 3
(3)		⟶	Amplified playback ACE head control pulse to pin 2
(4)		⟵	Capstan direction control
(5)	2v6		
(6)		⟶	Record mode to ACE control head
		⟵	Playback mode from ACE control head
(7)	0v		
(8)	2v7		
(9)		⟵	Record mode = H, Playback = L
(10)	2v8	⟵	Record mode vertical sync' for ACE head control pulse
(11)		⟵	5v0 Supply
(12)	2v3	⟶	5v pk-pk Pulse width modulated tracking control to servo
(13)		⟵	Tracking adjustment control
(14)	2v5	⟶	Head switching 5v pk-pk flip flop pulse
(15)	3v2	⟵	Head PG pulse adjustment
(16)	4v8	⟵	5v pk-pk drum motor PG pulses
(17)	3v9	⟵	5v pk-pk drum motor FG pulses
(18)	0v Ground		

Playback tape speed detection and Capstan gain compensation

Package type B (20 pins)

Pins		Function
(1)	←	Cue / review command = H
(2) 0v Ground		
(3) 0v Ground		
(4) 0v Ground		
(5)	←	5v Supply
(6)	→	Speed detected, LP mode = H
(7) n/c		
(8)	←	5v pk-pk speed & phase signal from Capstan servo
(9) 2v5		
(10) 2v6		
(11) 2v8	→	Capstan motor drive circuit
(12)		
(13)	←	LP = L, SP = H, from system control
(14) 5v0		
(15)	←	5v Supply
(16)	←	25 Hz Tracking control pulse from ACE head
(17)	→	Amplified & counted down Capstan FG pulses to Servo circuit
(18)	←	5v Supply
(19)	←	Capstan FG pulses
(20) 0v Ground		

Pins 6, 16,17 & 19 relevant to the tape speed detection section
Pins 1,8,10 & 11 relevant to the Capstan gain compensator section

AN6359

Part servo Capstan speed control

Package type B (20 pins)

Pins		Function
(1) 0v Ground		
(2) 0v Ground		
(3) 0v1	←	Fast speed = H, from system control
(4) 0v Ground		
(5)	←	5v Supply
(6) 0v		
(7) 5v0		
(8) 2v8	←	Capstan normal speed control from servo
(9) 2v8		
(10) 2v8		
(11) 2v6	→	Speed control to capstan motor drive circuit
(12)	←	5v Supply
(13) 2v6		
(14)	←	5v Supply
(15) 5v0		
(16)	←	ACE head control pulses
(17) 2v6	→	5v pk-pk squarewave capstan FG pulse ÷ 2, to servo
(18)	←	5v Supply
(19) 2v7	←	1v0 pk-pk capstan FG pulse
(20) 0v Ground		

AN6360

Record & playback colour processing **Package type B (18 pins)**

Pins		Function
(1)	←	Record mode 4.43MHz,200mV pk-pk chroma
(2) Automatic colour control circuit		
(3)	→	Playback mode burst pulse to external oscillator / sub convertor circuits
(4) 1v3	←	Burst gate pulse
(5)	←	9v Supply
(6)	←	Playback 4.43MHz chroma from pin 9 main via delay line
(7)	→	600mV pk-pk Playback 4.43MHz chroma to luminance / chroma mixer circuit
(8)	←	Record mode 140mV pk-pk, 627kHz chroma from pin 12
(9)	→	4.43MHz playback chroma to pin 6 via delay line
(10)	←	90mV pk-pk, 4.43MHz playback chroma from pin 12
(11) n/c		
(12)	→	Main convertor Record mode 627kHz chroma to pin 8 Playback mode 4.43MHz chroma to pin 10
(13)	←	Record mode 9v Supply
(14)	←	400mV pk-pk, 5.06MHz from external sub convertor circuit
(15) 0v Ground		
(16)	←	From pin 17
(17)	→	Playback 627kHz chroma to pin 16 Record 4.43MHz chroma to pin 16
(18)	←	Playback 627kHz from head preamplifier circuit

AN6367S

Record & playback Chroma processor

Package type B (22 pins)

Pins		Function
(1)	←	300mV pk-pk, 627kHz chroma from head amplifier
(2) 3v2		
(3) 3v2		
(4)	→	4.43 MHz to Servo
(5) 4.43 MHz xtal		
(6)	←	Burst gate pulse
(7)	→	Record 80mV pk-pk, Playback = 1v pk-pk, 5.06MHz to pin 18 via Band pass filter
(8)	←	5v Supply
(9) 4v0	←	Record & playback 627kHz, 4 × Phase shift signal
(10)	←	Pal = L, Secam = H control
(11)	→	Voltage controlled oscillator, 5.06MHz to external 4 × phase shifter circuit
(12)	←	Voltage controlled oscillator manual adjustment
(13)		
(14)	→	627kHz chroma, playback = 1v0 pk-pk, record = 1v2 pk-pk
(15)		
(16)	→	1v5 pk-pk, 4.43MHz playback chroma to pin 22
(17) 0v Ground		
(18)	←	From pin 7
(19)	←	Colour killer mode on = L
(20) 3v4, 0v5 pk-pk chroma		
(21)		
(22)	←	150mV pk-pk record 4.43MHz chroma E-E 600mV pk-pk playback

Drum motor drive

Package type B (24 pins)

Pins			Function
(1)	11v0	→	11v pk-pk drive to stator coil 1
(2)	11v0	→	11v pk-pk drive to stator coil 2
(3)	2v9		
(4)	2v6	→	Hall effect devices supply
(5)	0v1	←	Motor current feedback from pin 24
(6)	2v0	←	Motor on = H
(7)		←	12v Supply
(8)	0v1		
(9)	10v5		
(10)	1v0		
(11)	2v8		
(12)	0v Ground		
(13)	2v7	←	Drive from servo
(14)	0v7	←	–ve Hall effect device 1
(15)	0v7	←	+ve Hall effect device 1
(16)	1v4	←	–ve Hall effect device 2
(17)	1v4	←	+ve Hall effect device 2
(18)	3v0		
(19)	4v8	→	5v pk-pk PG pulse
(20)	3v8	→	5v pk-pk FG pulse
(21)		←	12v Supply
(22)	0v		
(23)	11v0	→	11v pk-pk drive to stator coil 3
(24)	0v1	→	To pin 5

AN6392

Video head record amplifier

Package type A (9 pins)

Pins			Function
(1)	4v4	←	Luminance FM & 625kHz record chroma input
(2)	6v0		
(3)	4v4		
(4)	4v4		
(5)	4v4 (half supply voltage)		
(6)	4v4		
(7)	0v Ground		
(8)	4v5	→	3v5 pk-pk record mode FM (SP) to head
(9)		←	9v Supply

AN6393

Playback FM agc & drop out detection — Package type B (18 pins)

Pins		Function
(1) 3v0		
(2) 0v Ground		
(3) 1v5		
(4) 4v0	→	FM automatic gain control to pin 14 (DOC) & pin 17
(5) 3v9	←	4v5 pk-pk Playback FM
(6) 4v2		
(7) 4v2	←	FM from pin 16 via High pass filter
(8) 4v2		
(9) 4v5		
(10) 3v5	←	700mV pk-pk FM from pin 16 via low pass filter
(11) 0v Ground		
(12) 3v2	→	Playback FM
(13)	←	5v Supply
(14) 3v2	←	From pin 4
(15)	→	Drop out compensator switching signal to external drop out phase noise cancelling circuits
(16) 2v3	→	To pin 10
(17) 3v3	←	From pin 4
(18)	←	5v Supply

AN6552

Dual operational amplifier

Package type B (8 pins)

Pins		Function
(1) 1v3	⟶	Amplifier 1, output
(2) 5v4	⟵	Amplifier 1, inverted input
(3) 3v5	⟵	Amplifier 1, non inverted input
(4) 0v Ground		
(5) 3v5	⟵	Amplifier 2, non inverted input
(6) 5v4	⟵	Amplifier 2, inverted input
(7) 1v3	⟶	Amplifier 2, output
(8)	⟵	12v Supply

Amplifiers designated numbers, i.e 1,2 are given to distinguish between the individual devices

AN6562

Dual operational amplifier

Package type B
(8 pins)

Pins		Function
(1)	→	Amplifier 1 output
(2)	←	Amplifier 1 inverted input
(3)	←	Amplifier 1 non inverted input
(4)	0v Ground	
(5)	←	Amplifier 2 inverted input
(6)	←	Amplifier 2 non inverted input
(7)	→	Amplifier 2 output
(8)	←	15v Supply.

Amplifiers designated numbers, i.e 1,2 are given to distinguish between the individual devices

AN6609N

Motor drive

Package type B (16 pins)

Pins		Function
(1)		
(2)		
(3)		
(4)		
(5)	←→	Motor drive -ve terminal
(6)		
(7) 0v Ground		
(8)	←→	Motor drive +ve terminal
(9)	←	14v Supply
(10)	←	14v Supply
(11)	←	14v Supply
(12) 0v Ground		
(13)	→	Voltage reference for pins 1,2, & 3
(14)	←	Control logic motor rotation
(15)	←	Control logic motor rotation
(16)	←	Control logic motor rotation

Drum motor drive

Package type B (24 pins)

Pins		Function
(1) 0v	←	–ve Hall effect device 1
(2) 2v0	←	+ve Hall effect device 1
(3) 0v	←	–ve Hall effect device 2
(4) 2v0	←	+ve Hall effect device 2
(5) 0v Ground		
(6) 11v3	→	Stator coil 1 & 2 drivers 6v2 supply, via external transistor
(7) 2v5	→	4v4 pk-pk stator coil 1
(8) 0v7		
(9) 0v6	←	Amplified drive from pin 20
(10) 6v2	←	Stator coil driver supply
(11) 2v2		
(12) 0v1		
(13) 2v1	→	Stator coil 2 drive
(14) 6v2	←	Stator coil driver supply
(15) 1v8		
(16) 1v8		
(17) 2v1	→	Stator coil 2 drive
(18)	←	Motor start control = H
(19)	←	5v Supply
(20)	←	Amplified motor drive to pin 9
(21) 2v8	←	Motor drive from servo
(22) 1v7	←	Hall effect device bias supply
(23) 3v7		
(24) 2v3	→	Hall effect device bias supply

AN6676

Capstan drive control

Package type B (24 pins)

Pins		Function
(1) 0v		
(2) 8v8		
(3)	←	Capstan stop = H
(4)	←	Capstan drive from servo
(5)	←	Torque control
(6) 0v		
(7)	←	Reverse brake = H
(8) 5v2		
(9) 5v9	←	1v7 pk-pk FG pulse 3 from motor
(10) 5v2		
(11) 5v2	←	1v7 pk-pk FG pulse 2 from motor
(12) 5v2		
(13) 0v Ground		
(14) 5v9	←	1v7 pk-pk FG pulse 1 from motor
(15) 0v, Oscillator pulses to FG coils		
(16) 5v8 Oscillator		
(17)	←	9v Supply
(18)	←	18v Supply
(19) 15v9	→	Motor drive 1 to external power transistor
(20) 15v8	→	Motor drive 2 to external power transistor
(21) 15v8	→	Motor drive 3 to external power transistor
(22) 0v3		
(23) 0v3		
(24)		

AN6677

Drum motor drive

Package type B (24 pins)

Pins			Function
(1)	4v5	←	Speed control from servo
(2)	5v1		
(3)		←	1v6 pk-pk, 200Hz Pulse from Drum PG head 1
(4)	5v1		
(5)		←	1v6 pk-pk, 200Hz Pulse from Drum PG head 2
(6)	5v2		
(7)		←	1v6 pk-pk, 200Hz Pulse from Drum PG head 3
(8)	5v2		
(9)	Oscillator		
(10)	0v Ground		
(11)	4v2	←	Drum on = H
(12)		←	9v Supply
(13)		→	FG pulse to Servo
(14)	0v2		
(15)	0v		
(16)	16v2		
(17)	1v4		
(18)		→	15v pk-pk Drum motor drive 1
(19)		→	15v pk-pk Drum motor drive 2
(20)	0v1 Three drive output common return to ground (via resistor)		
(21)		→	15v pk-pk Drum motor drive 3
(22)		←	Feedback from pin 20
(23)	8v0		
(24)		←	9v Supply

AN6873

8 × Buffer invertors **Package type B (18 pins)**

Pins		Function
(1)	←	Buffer 1 input
(2)	←	Buffer 2 input
(3)	←	Buffer 3 input
(4)	←	Buffer 4 input
(5)	←	Buffer 5 input
(6)	←	Buffer 6 input
(7)	←	Buffer 7 input
(8)	←	Buffer 8 input
(9)	←	–ve 20v Supply
(10) 0v Ground		
(11)	→	Buffer 8 output
(12)	→	Buffer 7 output
(13)	→	Buffer 6 output
(14)	→	Buffer 5 output
(15)	→	Buffer 4 output
(16)	→	Buffer 3 output
(17)	→	Buffer 2 output
(18)	→	Buffer 1 output

Uses include display segment drivers

Buffers designated numbers, i.e 1,2, etc are given to distinguish between the individual devices

AN6912

4 × Operational amplifiers

Package type B
(14 pins)

Pins		Function
(1)	→	Amplifier 1 output
(2)	→	Amplifier 2 output
(3)	←	12v Supply
(4)	←	Amplifier 2 inverted input
(5)	←	Amplifier 2 non inverted input
(6)	←	Amplifier 1 inverted input
(7)	←	Amplifier 1 non inverted input
(8)	←	Amplifier 4 inverted input
(9)	←	Amplifier 4 non inverted input
(10)	←	Amplifier 3 inverted input
(11)	←	Amplifier 3 non inverted input
(12) 0v Ground		
(13)	→	Amplifier 3 output
(14)	→	Amplifier 4 output

Amplifier designated numbers, i.e 1,2, etc are only given to distinguish between the individual devices

AN6913

Dual operational amplifiers

Package type A (9 pins)

Pins		Function
(1)	←	12v Supply
(2)	→	Amplifier 1 output
(3)	←	Amplifier 1 non inverted input
(4)	←	Amplifier 1 inverted input
(5) 0v Ground		
(6)	←	Amplifier 2 inverted input
(7)	←	Amplifier 2 non inverted input
(8)	←	Amplifier 2 output
(9)	←	12v Supply

Amplifiers designated numbers, i.e 1,2 are given to distinguish between the individual devices

AN6914

Dual comparator

Package type B (8 pins)

Pins		**Function**
(1)	⟶	Comparator 1 output
(2)	⟵	Comparator 1 inverted input
(3)	⟵	Comparator 1 non inverted input
(4)	0v Ground	
(5)	⟵	Comparator 2 non inverted input
(6)	⟵	Comparator 2 inverted input
(7)	⟶	Comparator 2 output
(8)	⟵	5v Supply

Comparators designated numbers, i.e 1,2 are given to distinguish between the individual devices

BA223

Video head horizontal switching pulse, 90 degree shift.
Long Play timing compensation

Package type A (8 pins)

Pins		Function
(1) 0v Ground		
(2)	←	LP head switching pulse
(3)	→	LP mode head switching pulse (90 degree shifted)
(4) n/c		
(5)	←	Phase / timing control
(6)	←	Temperature compensation circuit
(7)	←	LP mode, PG shift manual adjustment
(8)	←	9v Supply

BA3308

Dual audio amplifiers

Package type A (9 pins)

Pins		Function
(1)	←	Amplifier 1 inverted input
(2)	←	Amplifier 1 non inverted input
(3)	→	Amplifier 1 output
(4)	0v Ground	
(5)	←	Automatic level control
(6)	←	9v Supply
(7)	→	Amplifier 2 output
(8)	←	Amplifier 2 non inverted input
(9)	←	Amplifier 2 inverted input

Amplifiers designated numbers, i.e 1,2 are given to distinguish between the individual devices

BA340

Playback ACE control pulse amplifier

Package type A (8 pins)

Pins			Function
(1)	1v8, 2v5 pk-pk, control pulse waveform		
(2)	n/c		
(3)	0v5	←	Control pulse input from head
(4)	0v		
(5)	0v Ground		
(6)	4v4		
(7)		→	4v pk-pk control squarewave pulse output
(8)		←	5v Supply

BA5102

Audio record & playback amplifier

Package type B (18 pins)

Pins		**Function**
(1) 1v1		
(2) 0v5	←	Playback from audio head
(3) 0v		
(4) 3v8	→	Playback Audio to pin 8 via playback level control
	→	Record Audio to pin 9
(5) 0v Ground		
(6) 9v5		
(7) 0v4	←	E-E mode control
(8)	←	Playback audio from level control
(9)	←	Record audio from pin 4
(10) 4v6	→	Playback audio out to RF convertor & external line socket
	→	Record audio to head amplifier pin 12
(11) Record amplifier peaking circuit		
(12)	←	Record audio from pin10 via record level control
(13)	→	30v pk-pk Record mode audio to Head
(14)	←	Control, playback mode = Higher than 9v2, Record mode = Lower than 2v4, Mute mode = 2v4 to 9v2
(15)	←	12v Supply
(16)	←	Automatic loudness circuit
(17)		
(18) 0v6	←	Microphone input

BA5112LS

Audio record & playback amplifier **Package type B (24 pins)**

Pins		Function
(1)	←	12v Supply
(2)	←	Playback mute control
(3) 8v2	←	Record mute control
(4) 4v2		
(5) 0v		
(6) 4v1		
(7)	←	Record & playback audio from pin 13
(8) 4v2		
(9) 0v Ground		
(10)	→	Audio to record head
(11)		
(12)		
(13)	→	Playback & record audio to pin 7
(14)	←	Line audio input
(15) 3v6	←	Playback audio from pin 17
(16) 3v6		
(17)	→	Playback audio to level control and pin 15
(18)	←	Audio input from playback head
(19) 3v8		
(20) 8v7		
(21) 0v Ground		
(22) n/c		
(23) n/c		
(24) n/c		

Pins 22,23 & 24 contain a separate amplifier. Pins 23 & 24 are the inputs, pin 22 is the output

BA5115

Audio playback & record

Package type B (18 pins)

Pins		Function
(1) 5v1	←	Playback audio from pin 2 via level control
(2) 4v8	→	Playback audio to pin 1
(3) 5v5		
(4) 4v8	←	Playback from audio head
(5) 4v8		
(6)	←	12v Supply
(7) 5v5	←	3v0 pk-pk Record mode audio to head
(8) 4v8		
(9)	←	9v Supply, Record mode
(10)	←	Mute = 4v0
(11) 5v5	→	Audio output
(12) 0v Ground		
(13) 5v5		
(14) 3v0		
(15)		
(16) 4v		
(17)	←	Playback = H, Record = L
(18) 5v4	←	Record mode audio input

BA6109

Reel motor & Loading motor drive

Package type A (10 pins)

Pins		Function
(1) 0v Ground		
(2)	——→	Motor drive
(3)		
(4)	←——	Motor speed control
(5)	←——	Motor direction control input
(6)	←——	Motor direction control input
(7)	←——	15v Supply
(8)	←——	15v supply to reel / loading motor via output stage
(9)		
(10)	——→	Motor drive

Input pin 4 = H, Output pin 10 = L
Input pin 6 = H, Output pin 2 = L
Input pin 4 = L, Input pin 6 = H, Output pin 2 = L, Output pin 10 = H
Input pin 4 = H, Input pin 6 = L, Output pin 2 = H, Output pin 10 = L
Input pins 4 & 6 both L, Output pins 2 & 10 = Open

BA6121

Power supply control regulator

Package type B (16 pins)

Pins		**Function**
(1) 1v1	——→	Square wave drive
(2) 1v5		
(3) 0v Ground		
(4) 0v3		
(5) 3v0		
(6) 2v5, Triangle oscillator		
(7) 3v5		
(8) 7v4	——→	Square wave drive
(9) 3v0		
(10) 0v Ground		
(11) 2v4	——→	Square wave drive
(12)	←——	10v Supply
(13) 1v7		
(14) 3v6	——→	Square wave Drive
(15) 1v0		
(16) 3v0		

Squarewave drive outputs feed external power devices (transistors)

BA618

7 × Buffers

Package type B (16 pins)

Pins		Function
(1)	→	Buffer 1 output
(2)	→	Buffer 2 output
(3)	→	Buffer 3 output
(4)	→	Buffer 4 output
(5)	→	Buffer 5 output
(6)	→	Buffer 6 output
(7)	→	Buffer 7 output
(8)	←	12v Supply
(9) 0v Ground		
(10)	←	Buffer 7 input
(11)	←	Buffer 6 input
(12)	←	Buffer 5 input
(13)	←	Buffer 4 input
(14)	←	Buffer 3 input
(15)	←	Buffer 2 input
(16)	←	Buffer 1 input

Buffers designated numbers, i.e 1,2, etc are only given to distinguish between the individual devices

Carriage and/or loading motor drive

Package type A (10 pins)

Pins		Function
(1) 0v Ground		
(2)	→	Motor drive
(3)		
(4) n/c		
(5)	←	H = Pin 2 goes L L = Pin 2 goes H
(6)	←	H = Pin 10 goes L L = Pin 10 goes H
(7)	←	18v Supply
(8)	←	18v Supply
(9)		
(10)	→	Motor drive

BA6219

Capstan motor drive

Package type A
(10 pins)

Pins		Function
(1) 0v Ground		
(2)	⟶	Motor +ve terminal
(3)		
(4)	⟵	Mode control logic
(5)	⟵	Mode control logic
(6)	⟵	Mode control logic
(7)	⟵	18v Supply
(8)	⟵	14v Supply
(9)		
(10)	⟶	Motor –ve terminal

BA6222

Capstan motor control

Package type A (10 pins)

Pins		Function
(1)	0v Ground	
(2)	⟶	Capstan motor drive, Forward = H, Reverse = L Brake = L, Stop = Open
(3)		
(4)	⟵	Servo drive
(5)	⟵	Capstan drive control, Forward = H, Reverse = L Brake = H, Stop = L
(6)	⟵	Capstan drive control, Forward = L, Reverse = H Brake = H, Stop = L
(7)	⟵	12v Supply
(8)	⟵	12v Supply
(9)		
(10)	⟶	Capstan motor drive, Forward = L, Reverse = H Brake = L, Stop = Open

BA6229

Loading motor **Package type A (10 pins)**

Pins		Function
(1) 0v Ground		
(2) 0v5	——→	Loading motor drive, Loading = L, Unloading = H
(3) 0v		
(4) 0v	←——	Stop signal = L
(5) 0v	←——	Loading mode = L, Unloading mode = H, Stop = L
(6) 0v	←——	Loading mode = H, Unloading mode = L, Stop = L
(7)	←——	12v Supply to internal logic circuits
(8)	←——	15v Supply to internal motor driver
(9) 0v		
(10) 0v	——→	Loading motor drive, Loading = H, Unloading = L

Also used as a reel motor driver

Loading & Carriage control motor driver

Package type A (10 pins)

Pins		Function
(1) 0v Ground		
(2)	→	Loading mode. Reverse = H, Forward = L, Eject = open
(3)	→	Carriage motor. Reverse = open, Forward = open, Eject = H
(4)	←	Carriage motor control. Reverse = L, Forward = L, Eject = L Loading motor control. Reverse = H, Forward = H
(5)	←	Carriage motor control. Reverse = H, Forward = H, Eject = H Loading motor control. Reverse = L, Forward = L
(6)	←	Carriage motor control. Reverse = L, Forward = H, Eject = H Loading motor control. Reverse = L, Forward = H
(7)	←	12v Supply
(8)	←	Load & unload control
(9)	←	12v Supply
(10)	→	Carriage motor. Reverse = H, Forward = L Loading motor. Reverse = H, Forward = L Eject =L

BA6238A

Carriage & loading motor driver

Package type A (10 pins)

Pins		Function
(1)	0v Ground	
(2)	⟶	Loading motor –ve terminal
(3)	⟶	Carriage motor –ve terminal
(4)	⟵	Mode control logic
(5)	⟵	Mode control logic
(6)	⟵	Mode control logic
(7)	⟵	12v Supply
(8)	12v0	
(9)	12v0	
(10)	⟶	Loading motor & carriage motor +ve terminal

BA6259N

Carriage & loading motor control

Package type A (10 pins)

Pins		Function
(1) 0v Ground		
(2)	→	Carriage motor drive, Forward = L, Reverse = H Mode motor drive, Forward = Open, Reverse = Open Brake = L, Stop = Open
(3)	→	Carriage motor drive, Forward = Open, Reverse = Open Mode motor drive, Forward = L, Reverse = H Brake = L, Stop = Open
(4)	←	Carriage motor control, Forward = H, Reverse = H Mode motor control, Forward = L, Reverse = L Brake = H, Stop = H
(5)	←	Carriage motor control, Forward = L, Reverse = L Mode motor control, Forward = H, Reverse = H Brake = H, Stop = L
(6)	←	Carriage motor control, Forward = L, Reverse = H Mode motor control, Forward = L, Reverse = H
(7)	←	Motors 12v Supply source
(8) Connected to pin 9		
(9)	←	12v Supply
(10)	→	Carriage motor drive, Forward = H, Reverse = L Mode motor drive, Forward = H, Reverse = L Brake =L, Stop = Open

BA6301A

Drum servo

Package type B (16 pins)

Pins		Function
(1) 0v Ground		
(2) 1v4	←	+ve Drum motor FG pulse
(3) 1v4	←	–ve Drum motor FG pulse
(4) 1v4	→	Amplified Drum FG pulse
(5) 1v4	←	From pin 4
(6) 7v7	→	Squared FG pulses to pin 8 to
(7) 6v8	←	Drum on control
(8) 0v1	←	From pin 6
(9) 9v0	←	Trick speeds control
(10) 6v9		
(11) 7v5	→	Speed control signal to pin 13
(12) 4v0		
(13) 4v6	←	Phase control signal & input from pin 11
(14) 3v7		
(15) 0v7	→	Drum motor circuit drive
(16)	←	9v Supply

<u>Capstan servo</u>
Similar to above, except the capstan motor drive (error) signal exits at pin 11 as the speed (FG) portion. The phase signal requirement being added externally. Pins 12,13,14 & 15 not used.

BA6302A

Drum Discriminator

Package type B (16 pins)

Pins			Function
(1)	0v Ground		
(2)	1v4	←	+ve Drum motor FG pulse
(3)	1v4	←	–ve Drum motor FG pulse
(4)	1v4	→	Amplified Drum FG pulse
(5)	1v4	←	From pin 4
(6)	4v4	→	Squared pin 5 pulses to pin 11
(7)			
(8)	0v1		
(9)	9v0	←	Horizontal discriminator control
(10)			
(11)	4v4	←	From pin 6
(12)	2v6		
(13)	4v6	→	Frequency to voltage converted FG pulse to Drum phase i.c
(14)	4v7	←	From pin 13
(15)	4v6	→	Frequency to voltage converted FG pulse to phase comparator
(16)		←	9v Supply

BA6303

Capstan & Drum FG pulse amplifier & shaper

Package type B (16 pins)

Pins		Function
(1) 0v Ground		
(2) 2v5		
(3) 2v4	←	60mV pk-pk Capstan FG pulses
(4) 1v4	→	150mV pk-pk squarewave & amplified Capstan FG pulses
(5) 1v2	←	120mV pk-pk Drum FG pulse
(6) 2v4	→	500mV pk-pk squarewave amplified Drum FG pulse to pin 11
(7) n/c		
(8) n/c		
(9) n/c		
(10) n/c		
(11) 2v4	←	500mV pk-pk from pin 6
(12) 1v6	←	300mV pk-pk ramp waveform & Drum speed manual adjustment
(13) 2v8		
(14) 2v7		
(15) 2v1	→	Drum speed control output
(16)	←	5v Supply

There are two separate invertors connected between pins 7 & 10, and pins 8 & 9. The inputs are pins 7 & 8, outputs being 10 & 9 respectively.

BA6305

ACE head control pulse amplifier

Package type A (8 pins)

Pins			**Function**
(1)	1v4	←	Control pulse head
(2)	1v4		
(3)	1v4	→	1v pk-pk amplified control pulse
(4)	0v Ground		
(5)	2v0	←	From pin 3 to pulse squarer
(6)	2v7	→	4v pk-pk square wave control pulse to servo circuit
(7)		←	5v Supply
(8)		←	5v Supply

BA6328

Drum FG processor & 2 × invertors

Package type B (16 pins)

Pins			Function
(1)	0v Ground		
(2)		←	Drum FG pulse
(3)		←	Drum FG pulse
(4)		→	Amplified FG pulse to pin 5
(5)	0v5	←	From pin 4 to internal pulse squarer
(6)	1v4	→	Squarewave pulse output to pin 11
(7)		←	Pause & frame advance control
(8)		←	Pause control
(9)	3v0	→	Inverted pin 8 to capstan servo
(10)		→	Inverted pin 7 to capstan servo
(11)	1v5	←	From pin 6 to internal frequency to voltage convertor
(12)	2v0	←	Horizontal discriminator control
(13)	5v0		
(14)	4v7		
(15)	4v7	→	Converted FG frequency to voltage
(16)		←	9v Supply

BA634

Divide by two

Package type A (5 pins)

Pins		Function
(1)	←	4v6 pk-pk, 50Hz square wave pulse
(2)	→	4v0 pk-pk, 25Hz output
(3)	0v Ground	
(4)	←	9v Supply
(5)	←	Internal flip flop reset timing control pulse

BA6418N

Loading motor drive

Package type A (9 pins)

Pins		Function
(1)	←	Forward / reverse control
(2)	0v Ground	
(3)	←	Forward / reverse control
(4)	n/c	
(5)	0v Ground	
(6)	←	12v Supply
(7)	→	Motor drive
(8)	0v Ground	
(9)	→	Motor drive

BA6430S

Capstan motor drive

Package type B (24 pins)

Pins		Function
(1) 1v0	⟶	3v pk-pk stator coil 3 drive
(2) 0v2	⟶	Motor current sensing to pin 20
(3) 1v2	⟶	3v pk-pk stator coil 2 drive
(4) 0v Ground		
(5) 1v8	⟵	–ve Hall effect device 3
(6) 1v8	⟵	+ve Hall effect device 3
(7) 1v8	⟵	–ve Hall effect device 2
(8) 1v8	⟵	+ve Hall effect device 2
(9) 1v8	⟵	–ve Hall effect device 1
(10)	⟵	+ve Hall effect device 1
(11)	⟵	Motor Direction control, Reverse = H, Stop = Medium, Forward = L
(12) 0v Ground		
(13)	⟵	5v Supply
(14) 3v1	⟶	+ve Hall effect device bias
(15) 3v0	⟵	Free run manual adjustment
(16) 2v5	⟵	Capstan drive from servo
(17) 4v3		
(18) 4v1		
(19)	⟵	Motor stop control
(20) 0v2	⟵	From pin 2
(21) 0v Ground		
(22) 4v3		
(23) 1v5	⟶	3v pk-pk stator coil 1 drive
(24)	⟵	9v Supply

BA6432

Capstan motor drive

Package type B (24 pins)

Pins		Function
(1)	→	2v2 pk-pk drive to stator coil 3
(2)	→	Motor current torque sensing to pin 20
(3)	→	2v2 pk-pk drive to stator coil 2
(4) 0v Ground		
(5)	←	–ve from Hall effect device 3
(6)	←	+ve from Hall effect device 3
(7)	←	–ve from Hall effect device 2
(8)	←	+ve from Hall effect device 2
(9)	←	–ve from Hall effect device 1
(10)	←	+ve from Hall effect device 1
(11)	←	Motor control, Forward = L, Reverse = H, Stop = Medium
(12) 0v Ground		
(13)	←	5v Supply
(14)	→	Hall effect devices +ve supply
(15)	←	Free run adjustment
(16)	←	Drive control from servo
(17) n/c		
(18)		
(19)	→	Current sensing to servo
(20)	←	From pin 2
(21)		
(22)		
(23)	→	2v2 pk-pk drive to stator coil 1
(24)	←	10v Supply

BA6435S

Capstan motor drive

Package type B (24 pins)

Pins			Function
(1)	2v2	⟶	2v2 pk-pk drive to stator coil 3
(2)	0v1	⟶	Torque control to pin 20
(3)	2v3	⟶	2v2 pk-pk drive to stator coil 2
(4)	1v4		
(5)	2v2	⟵	–ve pulses from Hall effect device 3
(6)	2v2	⟵	+ve pulses from Hall effect device 3
(7)	2v2	⟵	–ve pulses from Hall effect device 2
(8)	2v2	⟵	+ve pulses from Hall effect device 2
(9)	2v2	⟵	–ve pulses from Hall effect device 1
(10)	2v2	⟵	+ve pulses from Hall effect device 1
(11)		⟵	Reverse, stop & forward control
(12)	0v Ground		
(13)		⟵	5v Supply
(14)	3v6	⟶	Hall effect devices bias supply
(15)	2v9		
(16)	2v6	⟵	Drive from servo
(17)		⟵	5v Supply
(18)	0v7		
(19)	0v5	⟵	Current limiting control
(20)	0v1	⟵	From pin 2
(21)	0v		
(22)	0v7		
(23)	2v3	⟶	2v2 pk-pk drive to stator coil 1
(24)		⟵	14v Supply

BA6459

Drum motor drive

Package type B (24 pins)

Pins			Function
(1)	2v3	←	Drive from servo
(2)		←	Start / stop 5v supply
(3)	4v0		
(4)	0v7		
(5)	2v8	→	Stator coil 1 drive
(6)		←	14v Supply
(7)	2v8	→	Stator coil 2 drive
(8)	0v		
(9)	2v8	→	Stator coil 3 drive
(10)	0v8		
(11)	2v2		
(12)	0v Ground		
(13)	2v5	←	+ve FG pulses
(14)	2v5	←	–ve FG pulses
(15)	2v5	→	FG pulses to servo circuit
(16)	2v5	→	PG pulses to servo circuit
(17)	2v5	←	–ve PG pulses
(18)	2v5	←	5v Supply
(19)	2v5	←	–ve pulses from Hall effect device 1
(20)	2v5	←	+ve pulses from Hall effect device 1
(21)	2v5	←	–ve pulses from Hall effect device 2
(22)	2v5	←	+ve pulses from Hall effect device 2
(23)	2v5	←	–ve pulses from Hall effect device 3
(24)	2v5	←	+ve pulses from Hall effect device 3

BA6871

Capstan motor drive

Package type B (32 pins)

Pins			**Function**
(1)		→	2v0 pk-pk stator coil 2 drive
(2)		←	14v Supply controlled by pin 4
(3)		→	2v0 pk-pk stator coil 1 drive
(4)		→	14v supply control circuit
(5)			
(6)		→	14v supply to external regulator
(7)	0v Ground		
(8)			
(9)	0v7		
(10)	2v8		
(11)		←	Capstan torque control
(12)	0v5		
(13)			
(14)		←	5v Supply
(15)	1v5		
(16)	0v Ground		
(17)	2v3	←	+ve Hall effect device 1
(18)	2v3	←	–ve Hall effect device 1
(19)	2v3	←	+ve Hall effect device 2
(20)	2v3	←	–ve Hall effect device 2
(21)	2v3	←	+ve Hall effect device 3
(22)	2v3	←	–ve Hall effect device 3
(23)	0v Ground		
(24)	6v3	←	FG pulse 1
(25)	6v3	→	FG pulse 1
(26)	6v3	←	FG pulse 2
(27)	6v3	→	FG pulse 2
(28)	6v3	←	12v Supply
(29)		←	Capstan motor direction control
(30)		→	2v0 pk-pk stator coil 3 drive
(31)		→	Output current sensing
(32)	0v Ground		

BA6993N

Dual operational amplifier **Package type A (8 pins)**

Pins		Function
(1)	⟶	Amplifier 1output
(2)	⟵	Amplifier 1 inverted input
(3)	⟵	Amplifier 1 non inverted input
(4)	0v Ground	
(5)	⟵	Amplifier 2 non inverted input
(6)	⟵	Amplifier 2 inverted input
(7)	⟶	Amplifier 2 output
(8)	⟵	9v Supply

BA7001

Video signal selector switch

Package type A (8 pins)

Pins			Function
(1)			
(2)		←	Input pin 7 or pin 8 select control
(3)	0v Ground		
(4)		→	Pin 7 or pin 8 output
(5)		→	Pin 7 or pin 8 output
(6)		←	9v Supply
(7)		←	Input 1
(8)		←	Input 2

BA7003

RF convertor

Package type A (16 pins)

Pins	Function
(1)	FM modulator LC circuit
(2)	FM modulator circuitry
(3)	FM modulator LC circuit
(4)	FM modulator circuitry
(5)	FM modulator circuitry
(6)	FM modulator circuitry
(7)	n/c
(8)	← Audio input
(9)	0v Ground
(10)	← Video input
(11)	→ Modulated video output
(12)	
(13)	
(14)	← Modulation oscillator bias adjustment
(15)	← 9v Supply
(16)	

BA7021

Signal switch

Package type A (9 pins)

Pins		Function
(1)	←	External video input 1
(2)	←	Input select control
(3)	←	External video. Input 2
(4)	←	Input select control
(5) 0v Ground		
(6)	←	Tuner. Input 3
(7)	←	Scart switch pin 8, playback = H or mute
(8)	→	Output from pin 3, 6 or 1
(9)	←	5v Supply

For clarity, typical signal sources and direction are shown. In practice one or more of the signals direction may differ.

BA7024

Video test signal generator **Package type A (7 pins)**

Pins		Function
(1)	←	5v Supply
(2) 0v, 500kHz xtal		
(3) 0v, 500kHz xtal		
(4) 0v Ground		
(5) 1v7	→	Composite video or test signal to RF convertor
(6)	←	Test signal = 12v, Video = 0v
(7) 3v0	←	Composite video

BA7131F

Video switching with mute facility

Package type B (8 pins)

Pins		Function
(1) 3v1	←	Video 1input
(2) 3v5	←	Output to pin 7 from pin 1 or pin 5 control
(3)		
(4) 0v Ground		
(5) 3v1	←	Video 2 input
(6)	←	Mute control
(7) 2v0	→	Video 1 or 2 output
(8)	←	5v Supply

Pin 6 may be fixed at 0v

BA718

Dual operational amplifiers

Package type A (9 pins)

Pins		Function
(1)	←	15v Supply
(2)	→	Amplifier 1 output
(3)	←	Amplifier 1, inverted input
(4)	←	Amplifier 1, non inverted input
(5)	0v Ground	
(6)	←	Amplifier 2, non inverted input
(7)	←	Amplifier 2, inverted input
(8)	→	Amplifier 2 output
(9)	←	15v Supply

Amplifier designated numbers, i.e 1,2, are only given to distinguish between the individual devices

BA7254

Video head amplifier

Package type B (32 pins)

Pins		Function
(1) 1v7	→	Playback Luminance output
(2) 3v1		
(3) 5v0	←	25Hz Head switching flip flop
(4) 0v9		
(5) 0v7	←	Playback FM from head 1
(6) 0v Ground		
(7) 0v7	←	Playback FM from head 2
(8) 0v8		
(9) 0v8		
(10) 0v7	←	'Trick mode playback head' 1
(11) 0v Ground		
(12) 0v7	←	'Trick mode playback head' 2
(13) 0v8		
(14)	←	Head amplifier 'normal or trick' select switching
(15) 0v Ground		
(16)	→	Record mode FM to heads
(17)		
(18)	→	Record mode FM to heads
(19)		
(20)		
(21)	←	Record mode 627kHz chroma input
(22)	←	Record mode Luminance input
(23)	←	12v Record mode supply
(24)		
(25) 2v4		
(26) 2v4		
(27) 4v7	→	Playback FM envelope
(28)	←	5v Playback mode Supply
(29) 3v0	→	200mV pk-pk playback chroma
(30) 5v0		
(31) 0v Ground		
(32) 2v6	→	Playback Luminance AGC output

BA7267

Colour processor **Package type B (22 pins)**

Pins		Function
(1) 4v8	←	400mV pk-pk 627kHz chroma from video head circuit
(2) 3v2		
(3) 3v3		
(4) 3v0, 4.43MHz xtal oscillator		
(5) 3v0, 4.43MHz xtal oscillator		
(6) 3v4	←	Burst pulse detection
(7) 3v5	→	Sub convertor 5.06MHz to main convertor pin 18
(8)	←	5v0 Supply
(9) 4v0	←	627kHz chroma sub convertor input from external 90 degree phase shifter circuit
(10) 1v8		
(11) 3v9	→	Voltage controlled oscillator output to external 90 degree phase shift circuit
(12)	←	Voltage controlled oscillator adjustment
(13) 2v5		
(14) 2v7	→	Main convertor output, Playback 4.43MHz to video output luminance / chroma mixer, Record 627kHz to Luminance / colour mixer video head amplifier
(15) 2v2		
(16) 1v8	→	Main convertor, 4.43MHz playback chroma to pin 22 via 2 line comb filter
(17) 0v Ground		
(18) 3v0	←	From pin 7
(19) 2v4	←	Colour on / off control
(20) 3v0	←	Record 4.43MHz or playback 627kHz chroma to main convertor from pin 21
(21) 3v4	→	To pin 20
(22) 3v0	←	4.43MHz chroma, record mode input from tuner or playback from pin 16

BA7740FS

Drum mounted audio heads playback & record pre-amplifier

Package type B (24 pins)

Pins		Function
(1)	←	Record mode 5v Supply
(2) 0v6		
(3) 2v8	←	Record mode audio input
(4) 1v3		
(5)	←	Playback mode = H, record mode = L
(6)	←	5v Supply
(7) 2v9	→	Audio RF output
(8) 2v5		
(9) 2v5		
(10) n/c		
(11) 0v		
(12) 5v0	←	Audio head switching flip flop pulse
(13)	←	5v Supply
(14) Record mode = 5v0		
(15) Record mode = 5v0		
(16) 0v		
(17)	←	Playback from audio head 1
(18) 0v Ground		
(19)	←	Playback from audio head 2
(20)		
(21) 0v Ground		
(22)	→	Record mode to heads
(23) n/c		
(24) 300mV pk-pk record mode test output point		

BA7751AL

Audio record & playback

Package type A (24 pins)

Pins		Function
(1)	←	Playback from audio head
(2)		
(3)		
(4)		
(5)		
(6)	←	12v Supply
(7)	→	Amplified playback to pin 9 via playback level control
(8)	←	Tuner or external audio
(9)	←	From pin 7
(10) 5v7		
(11)		
(12) 5v7		
(13)		
(14) 4v5		
(15)	←	Audio mute control
(16)	→	2v pk-pk audio to RF convertor & line out
(17)	→	Playback / record head switching circuit
(18)	←	Playback = H, Tuner = L
(19)	→	3v0 pk-pk Record mode audio to head
(20)	←	Record = L, Playback = H
(21) 5v7		
(22) 0v		
(23)		
(24) 0v Ground		

BA7752LS

Audio playback & record

Package type A (24 pins)

Pins		Function
(1) 5v6	←	Playback audio from head
(2) n/c		
(3) 5v6		
(4) 12v1		
(5) 5v5		
(6)	←	12v Supply
(7) 5v5	→	Amplified playback to pin 9
(8) 6v1	←	Record mode audio input
(9) 6v1	←	From pin 7
(10) 6v2		
(11) 6v1		
(12) 6v2		
(13)		
(14) 5v5		
(15)	←	Audio mute = H
(16) 6v1	→	Record & playback audio output
(17) 11v5		
(18)	←	E-E mode = H
(19) 6v1	→	Record mode audio to head
(20)	←	Record = H
(21) 6v2		
(22)		
(23)		
(24) 0v Ground		

BA7755

Single pole make/break switch

Package type A (5 pins)

Pins		Function
(1)	←	12v Supply
(2)	→	Switch A terminal
(3)	←	Switch Pole terminal
(4)	←	Logic control
(5)	←	Logic control

For clarity the signal is shown entering via the pole terminal and exit at A terminal. In practice the signal flow could be in any direction.

BA7765

Audio record & playback amplifier

Package type B (32 pins)

Pins		Function
(1) 0v Ground		
(2)	←	Record / playback head switch control
(3) 0v		
(4) 2v0	←	Playback audio
(5) 2v0		
(6) 2v0		
(7)		
(8) 2v0	→	Playback audio
(9) 0v	←	External or E-E audio
(10) 5v0	←	Playback or E-E select control
(11) 0v	←	Line or external audio
(12) 0v		
(13) 6v5	←	Automatic level control
(14) 4v1	→	Line or external audio
(15) 4v2	←	Audio record from pin 14
(16) 4v2	←	Audio playback from pin 14
(17) 4v2	→	Audio out to pin 20 and external
(18)	←	Mute control
(19) 0v		
(20) 0v	←	Record audio from pin 17
(21) 0v		
(22) 0v		
(23) 0v4		
(24) 4v0	→	Record audio to head
(25)	←	Record or E-E control
(26) 0v		
(27) 0v1		
(28)	←	9v Record supply
(29)	←	9v Supply
(30) 8v8		
(31) 8v0		
(32)	←	Record head switching

Pin 2 may not be connected

BA7766AS

Audio playback & record

Package type B (32 pins)

Pins		**Function**
(1) 0v Ground		
(2)		
(3) 0v		
(4) 2v0	←	Playback from audio head
(5) 2v0	←	From pin 6
(6) 2v0	→	Switched PB feedback to pin 5
(7) 2v0		
(8) 2v0	→	Playback to pin 9 via level control
(9)	←	From pin 8
(10)	←	Record = H, Playback = L
(11) 0v	←	Record mode audio input
(12) 0v		
(13)	←	10v Supply from pin 30
(14) 6v0	→	E-E & playback audio output & to pin 20
(15) 6v0		
(16) 5v6		
(17) 6v0		
(18)	←	Mute = H
(19)	←	LP mode = H
(20)	←	Record mode input
(21) 0v Ground in LP mode		
(22) 0v Ground in LP mode		
(23)		
(24)	→	Amplified record mode audio to pin 32
(25)	←	0v Record mode, 9v playback mode
(26)	←	12v Record mode, 0v playback mode
(27)	→	11v record mode supply from internal regulator to Bias oscillator
(28)	←	12v Record mode supply to internal regulator
(29)	←	12v Supply
(30) 12v0		
(31)	→	10v regulated supply to pin 13

BA7790LS

Audio record & playback

Package type B (24 pins)

Pins		Function
(1)	0v Gropund, LP mode	
(2)	←	Playback audio from head
(3)	Ground in Playback mode	
(4)	0v Ground	
(5)		
(6)		
(7)	←	9v Supply
(8)		
(9)	←	Mode control LP = H, SP = L
(10)	→	Playback audio to pin 12
(11)	←	E-E mode = H
(12)	←	Playback audio from pin 10
(13)		
(14)		
(15)		
(16)	←	External audio input
(17)	←	Mute = H
(18)	←	Audio from tuner
(19)		
(21)	→	Record mode audio to head
(22)	→	Playback audio
(23)	0v Ground in LP mode	
(24)	←	Record mode = H, Playback = L

BA833S

Capstan & Drum servo

Package type B (42 pins)

Pins		Function
(1) n/c		
(2)	←	Record = H, Playback = L
(3)	←	Record = H, Playback = L
(4)	←	Capstan phase control from micro'
(5)	←	Drum phase control from micro'
(6)	←	5v Supply
(7) 4v6		
(8) 4v6		
(9) 4v6		
(10) n/c		
(11) n/c		
(12)	←	Drum flip flop pulse
(13)		
(14) 4v6		
(15) 4v4	→	Drum motor drive circuits
(16) 0v5		
(17) 1v3, 32.7kHz xtal		
(18) 2v8, 32.7kHz xtal		
(19)	→	Record mode control pulses to ACE head
(20)		
(21) 0v Ground		
(22)		
(23)		
(24)	←	Vertical sync' from composite video
(25)	←	ACE head control pulse
(26) 4v2	←	Capstan motor FG pulse
(27)		
(28)		
(29) n/c		
(30) n/c		
(31)		
(32)	←	Horizontal discriminator control from micro'
(33)		
(34)	→	Capstan motor drive circuits
(35) 4v6		
(36)		
(37)		

(Continued opposite)

(38) ⟵ Tracking control
(39)
(40) 0v Ground
(41) ⟵ 9v Supply
(42) n/c

BA874L

Reel motor control **Package type B (16 pins)**

Pins		Function
(1)	←	Slow & still mode control
(2)	←	Rewind & fast forward = H
(3)	←	Play, forward search & review = H
(4)	←	Play, stop, forward search & review = H
(5)	→	Forward search = L
(6)	→	Review = L
(7)		
(8)	→	Reel motor drive transistor
(9)	→	Reel motor drive transistor
(10)	→	Forward search & review = H
(11)	→	Forward search & rewind = H
(12)	→	Play, load & unload = H
(13)	→	Play, load & unload = H
(14)	←	LP mode = L
(15) 0v Ground		
(16)	←	12v Supply

BAF6305

ACE head control pulse amplifier & squarer

Package type B (8 pins)

Pins			**Function**
(1)		←	Control pulse from ACE head
(2)	Inverted input		
(3)	1v3	→	800mV pk-pk amplified control pulses to pin 5
(4)	0v Ground		
(5)		←	From pin 3
(6)	1v3	→	Squared control pulses to servo circuit
(7)	n/c		
(8)		←	5v Supply

BAL6309

Artificial vertical sync' generator

Package type A (16 pins)

Pins		Function
(1) 0v Ground		
(2)		
(3) 4v1		
(4) 0v1		
(5)	⟶	Tape speed control, SP = H
(6)	⟶	Tape speed control, LP = H
(7)	⟵	Speed mode control, LP = H
(8)	⟵	Vertical sync' pulse manual width adjustment
(9)	⟵	5v Supply
(10)	⟶	10v pk-pk Artificial vertical sync' pulse to Video circuits
(11) 5v0		
(12) n/c		
(13) 5v1		
(14)	⟵	Trick playback mode control, Normal = H
(15)	⟵	Trick playback mode control, Normal = L
(16) 2v3	⟵	Head switching pulse

Prevents vertical jitter during playback trick modes.

BN5111B

Vision I.F amplifier & demodulator

Package type B (28 pins)

Pins		Function
(1) 4v6	←	Vision I.F input
(2) 0v Ground		
(3) 4v5		
(4) 3v8		
(5) 10v5	→	Automatic gain control (AGC) to tuner
(6) 4v2	←	AGC control
(7) 3v6	→	Vision I.F to pin 9
(8) 0v Ground		
(9)	←	From pin 7
(10) 6v6	→	Automatic frequency control (AFC) to Tuner
(11) 2v8, AFC detector LC circuit		
(12) 6v2, Vision detector LC circuit		
(13) 6v2, Vision detector LC circuit		
(14) 2v8, AFC detector LC circuit		
(15) 2v1		
(16) 0v Ground		
(17) 4v3	→	Video to pin 18
(18) 4v3	←	Video from pin 17
(19) 6v8	→	Video output
(20)	←	12v Supply
(21) 5v3	→	Sound I.F output
(22) 0v Ground		
(23) 5v0		
(24)		
(25) 5v0		
(26) 4v5		
(27) 9v0		
(28) 4v5	←	Vision I.F input

BN5115B

Vision & sound I.F amplifier & detector

Package type B (22 pins)

Pins		Function
(1)	←	12v Supply
(2) 6v5	→	Video output
(3) 9v8		
(4) 0v Ground		
(5) 5v0		
(6) 5v0	←	I.F input
(7) 5v0	←	I.F input
(8) 5v0		
(9) n/c		
(10) 10v7	→	Tuner AGC control
(11) 4v2	←	AGC manual adjustment
(12) 4v5	→	Amplified I.F to pin 14
(13) 10v2	←	AFC on / off control
(14) 5v4	←	From pin 12
(15) 6v6	→	Automatic frequency control (AFC) to tuner
(16) 3v0		
(17) 7v5, Video detector LC circuit		
(18) 7v5, Video detector LC circuit		
(19) 3v0, AFC detector LC circuit		
(20) 6v4	→	Sound I.F output
(21) 5v5	→	Detected video to pin 22 via filter
(22) 5v5	←	From pin 21

BU2735

Drum & Capstan digital servo

Package type B (30 pins)

Pins			Function
(1)	2v5	←	1v0 pk-pk, 4.43MHz pulse
(2)	0v Ground		
(3)	2v5	→	Drum motor speed control
(4)	2v5	→	Drum motor phase control
(5)	0v Ground		
(6)	2v5	←	4v0 pk-pk Drum FG pulse
(7)	3v6	←	2v0 pk-pk Drum PG pulse
(8)	0v5		
(9)	2v3	←	Head switching pulse
(10)			
(11)	0v1	←	Head switching point vertical control
(12)		→	Vertical sync' pulse
(13)	0v1	←	Enable
(14)	0v1	←	Data
(15)	0v1	←	Clock pulse
(16)	0v5	←	Composite sync'
(17)		→	Reference 25Hz
(18)		→	Capstan rotation duty cycle sensor to micro'
(19)		→	Capstan speed SP = H, LP = L
(20)		→	Capstan speed SP = L, LP = H
(21)	2v6	←	Capstan FG pulse
(22)	2v6	←	2v0 pk-pk capstan PG pulse
(23)	0v1	←	Record control pulse control
(24)	1v7		
(25)			
(26)	2v5	→	4v0 pk-pk Capstan FG pulse to micro'
(27)	2v5	→	ACE head record mode 5v pk-pk control pulse
(28)		←	5v Supply
(29)	2v6	→	4v0 pk-pk capstan phase control
(30)	2v6	→	4v0 pk-pk capstan speed control

BU2745

Drum & Capstan servo — **Package type B (30 pins)**

Pins			Function
(1)	2v6	←	4.43MHz, 700mV pk-pk clock pulse
(2)	0v Ground		
(3)	2v6	→	5v0 pk-pk squarewave Drum speed control output
(4)	2v5	→	5v0 pk-pk squarewave Drum phase control output
(5)	5v0		
(6)	2v6	←	Record & playback 4v0 pk-pk Drum FG pulses
(7)	2v4	←	Record & playback 4v0 pk-pk Drum Tacho square-wave pulses
(8)	0v7	←	Head switching adjustment
(9)	3v0	→	25Hz flip flop switching pulse
(10)	2v3		
(11)	0v5	←	Vertical lock control
(12)	5v0		
(13)		←	Serial data
(14)		←	Serial data
(15)		←	Serial clock
(16)	0v4	←	Record & playback Vertical sync' pulse
(17)	0v4	→	Vertical sync' pulse out
(18)	5v0	→	SP = H, LP = L
(19)	5v0	→	LP = L
(20)	0v2		
(21)	2v5	←	4v0 pk-pk Capstan FG pulses
(22)	2v8	←	4v0 pk-pk squarewave ACE head control pulses
(23)	0v		
(24)	2v7	←	Tracking control
(25)	0v	→	Control pulse detector output
(26)	5v0		
(27)	3v5	→	5v0 pk-pk record control pulses to ACE head
(28)		←	5v0 Supply
(29)	2v7	→	Capstan phase control
(30)	2v6	→	Capstan speed control

BU2751S

Drum & Capstan servo processor

Package type B (30 pins)

Pins		Function
(1) 2v4	←	4.43MHz
(2) 0v Ground		
(3) 2v4	→	Drum speed control
(4) 2v4	→	Drum phase control
(5) 0v Ground		
(6) 2v4	←	Drum FG pulse
(7) 4v2	←	Drum PG pulse
(8) 0v5	←	Head switching adjustment
(9) 2v6	→	Head switching pulse
(10) 2v5		
(11) 0v5	←	Vertical pulse
(12) 0v	→	Vertical pulse
(13) 2v7	←	Data load
(14) 0v5	←	Data
(15) 1v3	←	Clock pulse
(16) 0v5	←	Composite sync' pulse
(17)		
(18) 0v	→	Control duty information to system control micro'
(19) 4v9	→	Capstan speed data to system control micro'
(20)		
(21) 2v2	←	Capstan FG pulse
(22) 2v8	←	Record mode control
(23) 0v	←	ACE head control pulse
(24) 1v8	←	Tracking control
(25) 4v9		
(26) 4v9	→	Capstan FG pulse
(27) 0v2	→	Record mode ACE control pulse
(28)	←	5v Supply
(29) 2v4	→	Capstan motor phase control
(30) 2v4	→	Capstan motor speed control

BU2890BK

Digital servo processor

Package type C (44 pins)

Pins		Function
(1)	——→	Capstan speed control to pin 33
(2) 0v Ground		
(3)	←——	4.43MHz clock
(4)	——→	LP mode = H
(5)	——→	n/c or Head switching flip flop pulse to chroma circuit.
(6)	←——	Data from system control
(7)	←——	Clock from system control
(8)	←——	Head switching point control
(9)	——→	Head switching pulse
(10) n/c		
(11)	←——	Composite sync' pulses
(12)	——→	Picture vertical lock pulse
(13)	——→	Capstan motor forward / reverse control
(14)	——→	Capstan FG pulse to system control
(15)	——→	ACE head control pulse to system control
(16)	←——	Capstan motor brake control
(17) n/c		
(18)	——→	ACE head Control pulse data sensing to system control
(19)	——→	Tape speed detection to system control
(20) 2v4		
(21)	←——→	Record & playback to / from control head
(22)	←——→	Record & playback to / from control head
(23)	←——	5v Supply
(24) 0v Ground		
(25)		
(26)		
(27)		
(28)	←——	Capstan FG pulse
(29)	←——	Drum FG pulse
(30)	←——	Drum PG pulse
(31)	——→	Capstan motor drive
(32)	←——	Capstan digital to analogue converted phase control from pin 44
(33)	←——	Capstan digital to analogue converted speed control from pin 1

(Continued opposite)

(34) → Head drum motor drive
(35) ← Drum digital to analogue converted phase control from pin 41
(36) ← Drum digital to analogue converted speed control from pin 42
(37) ← Drum control, slow & still modes
(38) ← FM Envelope from head pre-amplifier
(39) n/c
(40) → Head switching flip flop to head pre-amplifier
(41) → Drum phase control to pin 35
(42) → Drum speed control to pin 36
(43) ← 5v Supply
(44) → Capstan phase control to pin 32

BX6385

Video mute

Package type A (10 pins)

Pins			Function
(1)	2v0	←	Video input
(2)	0v Ground		
(3)	2v5		
(4)	1v3	→	Video to pin 6
(5)	0v Ground		
(6)	1v3	←	Video from pin 4
(7)	1v9	→	Normal or muted video output
(8)			
(9)	0v	←	Picture mute control = H
(10)		←	5v Supply

CD4053

3 × 2 way single pole switches **Package type B (16 pins)**

Pins		Function
(1)	——→	Switch 1 terminal A
(2)	——→	Switch 1 terminal B
(3)	——→	Switch 2 terminal B
(4)	←——	Switch 2 pole terminal
(5)	——→	Switch 2 terminal A
(6)	0v Ground	
(7)	←——	5v Supply reset
(8)	0v Ground	
(9)	←——	Switch 2 control
(10)	←——	Switch 1 control
(11)	←——	Switch 3 control
(12)	——→	Switch 3 terminal A
(13)	——→	Switch 3 terminal B
(14)	←——	Switch 3 pole terminal
(15)	←——	Switch 1 pole terminal
(16)	←——	5v Supply

Switch designated numbers, i.e 1,2, etc are only given to distinguish between the individual devices
For clarity the signal is shown entering via the pole terminal and exit at either A or B terminal. In practice the signal flow could be in any direction.

CD4066

4 × single pole, one way switches

Package type B (14 pins)

Pins		Function
(1)	←	Switch 1 pole
(2)	→	Switch 1 closed or open
(3)	←	Switch 2 pole
(4)	→	Switch 2 closed or open
(5)	←	Switch 2 open / closed control
(6)	←	Switch 3 open / closed control
(7)	0v Ground	
(8)	→	Switch 3 closed or open
(9)	←	Switch 3 pole
(10)	→	Switch 4 closed or open
(11)	←	Switch 4 pole
(12)	←	Switch 4 open / closed control
(13)	←	Switch 1 open / closed control
(14)	←	12v Supply

Switch designated numbers, i.e 1,2, etc are only given to distinguish between the individual devices
For clarity the signal is shown entering the switches via their pole terminals. In practice the signal direction could be the reverse.

CXK1006

Non volatile RAM memory

Package type A (10 pins)

Pins		Function
(1)	Clock RC circuit, erase / write timing	
(2)	←	Clock pulse
(3)	←	Enable = L
(4)	↔	Address & data
(5)	↔	Address & data
(6)	↔	Address & data
(7)	↔	Address & data
(8)	0v Ground	
(9)	5v	
(10)	←	5v Supply

CXK1008

Non volatile Ram memory

Package type A
(10 pins)

Pins		Function
(1)	RC Oscillator circuit	
(2)	←——	Clock pulse
(3)	←——	Reset
(4)	←——→	Address & data
(5)	←——→	Address & data
(6)	←——→	Address & data
(7)	←——→	Address & data
(8)	0v Ground	
(9)		
(10)	←——	5v Supply

Loading Motor control

Package type B (10 pins)

Pins		Function
(1) 0v Ground		
(2)	——→	Loading motor drive –ve
(3)		
(4)	←——	Motor torque control
(5)	←——	Motor control (+ve)
(6)	←——	Motor control (–ve)
(7)	←——	12v Supply
(8)	←——	12v Supply
(9)		
(10)	——→	Loading motor drive +ve

HA11107

Sound FM detector

Package type B (14 pins)

Pins		Function
(1) 1v8		
(2) 1v8	←	Sound I.F input
(3) 0v Ground		
(4) 0v Ground		
(5)	←	10v Supply
(6) 1v7	←	Sound mute control
(7) 6v5		
(8) 5v5	→	Audio output
(9) FM detector LC circuit		
(10) FM detector LC circuit		
(11) n/c		
(12)	→	Separate audio amplifier 1 output
(13)	←	Separate audio amplifier 1 input
(14)	←	Separate audio amplifier 1 input

Pins 12,13 & 14 are connected to an independent amplifier within the device. When not used pins 13 & 14 may be connected to 0v Ground.

HA11215A

Sound & vision I.F demodulator

Package type B (24 pins)

Pins		Function
(1) 5v7		
(2) 5v0	⟶	Composite video to pin 3 & Sound I.F circuit
(3)	⟵	Video from pin 2
(4)	⟵	Video level adjustment
(5) 4v5		
(6) 3v6		
(7) 3v4	⟵	I.F from SAW Filter
(8) 4v5		
(9) 7v0		
(10)	⟶	Automatic gain control (AGC)to tuner
(11) 0v Ground		
(12)	⟵	AGC level control
(13) 4v5		
(14) 4v5		
(15)		
(16)		
(17) 4v2		
(18) 6v6 Video demodulator LC circuit		
(19) 6v6 Video demodulator LC circuit		
(20)		
(21)	⟵	12 Supply
(22)		
(23)		
(24)	⟶	Composite video output

HA1124A

Sound FM detector

Package type B (14 pins)

Pins		Function
(1) 1v8		
(2) 1v8	←	Sound I.F input
(3) 0v Ground		
(4) 0v Ground		
(5)	←	10v Supply
(6) 1v7	←	Sound mute control
(7) 6v5		
(8) 5v5	→	Audio output
(9) FM detector LC circuit		
(10) FM detector LC circuit		
(11) n/c		
(12)	→	Separate audio amplifier 1 output
(13)	←	Separate audio amplifier 1 input
(14)	←	Separate audio amplifier 1 input

Pins 12,13 & 14 are connected to an independent amplifier within the device. When not used pins 13 & 14 may be connected to 0v Ground.

HA11711

Drum servo

Package type B (28 pins)

Pins		**Function**
(1) 0v5, 32 Khz xtal oscillator linked to pin 28		
(2)	←	25Hz, 1v6 pk-pk amplified playback control pulses from ACE head
(3) n/c		
(4)	→	Record mode, control pulses to ACE head
(5)	→	Regulated 6v3
(6) n/c		
(7) 0v Ground		
(8)	←	Record mode 500mV pk-pk vertical sync'
(9)	←	12v Supply
(10)		
(11)		
(12) 6v1	→	Drum phase error signal to drive circuit
(13) 5v6		
(14) 7v7		
(15) 1v1		
(16) 6v0	→	Drum flip flop to chroma circuit
(17) 0v4	←	Head 1 phase control
(18) 0v4	←	Head 2 phase control
(19)	←	600mV pk-pk drum FG pulse
(20)	←	Record mode 12v
(21) 0v Ground		
(22) 3v0	←	Record mode, capstan phase servo comparison signal
(23) 6v4 connected to pin 5		
(24) 7v3		
(25) 6v5	→	Capstan phase error signal to drive circuit
(26)		
(27)		
(28) 32 Khz xtal oscillator linked to pin 1		

HA11724

Record & playback Luminance processor

Package type B (28 pins)

Pins		Function
(1)	←	Record FM & 627Khz chroma
(2)		
(3)	→	300mV pk-pk Record FM to video heads
(4)		
(5)	←	Playback FM from video head 2
(6) 0v Ground		
(7)	←	Playback FM from video head 1
(8)		
(9)	←	Playback 9v supply
(10) Record FM modulator circuit		
(11) Record FM modulator circuit		
(12)	→	Record mode Luminance to pin 17 via de-emphasis circuit
(13)		
(14)	←	Dark clip control
(15)	←	White clip control
(16)		
(17)	←	From pin 12
(18)	←	Record mode Luminance
(19) Record mode non linear pre-emphasis circuit		
(20)	←	25Hz head switching flip flop pulse
(21) 0v Ground		
(22)	→	500mV pk-pk Playback FM to external demodulator circuit & pin 23 DOC via delay line
(23)	←	Delayed playback FM from pin 22
(24)	→	Playback demodulated video to external DOC circuit
(25)		
(26)	←	Playback 9v supply
(27)	←	Playback FM from pin 28 via high pass filter
(28)	→	Amplified video heads FM to pin 27

HA11738

Record & playback luminance processor

Package type B (28 pins)

Pins		Function
(1) 1v9		
(2)		
(3) 350mV pk-pk Horizontal sync' pulse		
(4)		
(5)	→	E-E or Playback composite video to RF out
(6)	←	Vertical sync' pulse from during still & search modes
(7)	←	Playback 4.43MHz chroma
(8)	←	E-E level control
(9)		
(10)	←	Playback mode 9v supply
(11)		
(12)	←	Playback luminance via level control & demodulator
(13)	←	Record mode luminance from pin 24 via 4.43MHz chroma trap
(14)	→	Record mode Luminance to pin 18 FM modulator, colour AFC circuit & AGC circuit pin 28. Playback mode Luminance to pin 28 & colour AFC circuit
(15) 4v2		
(16)	←	Playback FM input
(17) 0v Ground		
(18)		
(19)		
(20)	←	9v Supply
(21)		
(22)	←	Playback 9v supply
(23)		
(24)	→	Demodulated playback Luminance to pin 12
(25)		
(26)	←	Record mode Luminance from Tuner or external input
(27)		
(28)	←	Luminance from pin 14

HA11744

Video head record & playback processor

Package type B (28 pins)

Pins		Function
(1)	←	Record FM luminance input
(2)	←	Record 627Khz chroma input
(3)	→	Record 627Khz chroma mode to heads
(4)		
(5)	←	Playback FM from head 2
(6) 0v Ground		
(7)	←	Playback FM from head 1
(8)		
(9)	←	Playback mode 9v Supply
(10) 7v4 FM modulator circut		
(11) 7v4 FM modulator circuit		
(12)	←	Dark clip adjustment
(13)	←	FM modulator deviation control
(14)	←	9v Supply
(15)	←	25Hz flip flop switching pulse
(16)	←	White clip adjustment
(17) 3v2		
(18) 3v2		
(19)	→	Luminance FM output
(20)		
(21) 0v Ground		
(22)	→	FM Luminance to drop out compensation delay line
(23)		
(24)		
(25) 7v0		
(26)	←	Luminance AGC control
(27)		
(28)	→	627Khz playback chroma output

HA11751

Drum & Capstan Servo

Package type B (42 pins)

Pins			Function
(1)			
(2)	4v5	→	7v pk-pk, 25Hz Drum flip flop pulse
(3)		→	Vertical pulse, Still & Search mode
(4)	0v Ground		
(5)			
(6)	4v5	→	Drum motor phase control
(7)		→	Record mode, control pulses to ACE head
(8)		←	Forward & reverse search mode control
(9)		←	Vertical sync' pulse from video
(10)		←	9v Supply
(11)		←	25Hz control pulse from ACE head amplifier
(12)		←	Mode control, Record = H, PB = L
(13)		←	Mode control, Record = H, PB = L
(14)		←	9v Supply
(15)		←	Capstan FG Pulse
(16)	4v3		
(17)	0v		
(18)		←	Search mode = H
(19)	0v Ground		
(20)		←	Still mode = H
(21)	0v		
(22)			
(23)			
(24)			
(25)			
(26)		←	Stop, forward, reverse search, still & unloading modes = H
(27)		→	Capstan motor phase control
(28)			
(29)		←	Manual tracking control
(30)	4.43MHz xtal		
(31)	4.43MHz xtal		
(32)		←	5v Supply
(33)		→	To pin 34
(34)		←	From pin 33
(35)			
(36)		←	Vertical lock control

(Continued overleaf)

HA11751

(37)
(38) ← 2v pk-pk, 25Hz Flip flop pulses
(39) 0v Ground
(40) ← Video head 2 manual switching control
(41) ← Video head 1 manual switching control
(42)

HA11756

Colour processor

Package type B (28 pins)

Pins		Function
(1) 3v1	←	chroma from delay line
(2) 3v3	→	Record mode 625kHz chroma
(3) 3v2	→	Playback 625kHz chroma to delay line
(4) 0v8		
(5) 3v8	←	160mV pk-pk, Playback 625kHz chroma from pin 23
(6) 0v Ground		
(7) 1v0		
(8) 0v	←	Composite sync' pulses
(9) 2v6		
(10) 3v7	→	5.06MHz, 800mV pk-pk playback & record to pin 27
(11) 2v9		
(12) 2v2		
(13) 2v7		
(14) 2v2		
(15) 2v7, 4.43MHz xtal		
(16) 3v0, 4.43MHz xtal		
(17) 4v0		
(18) 2v9		
(19) 2v8	←	Horizontal sync' pulses
(20) 3v0		
(21) 4v1	→	600mV pk-pk, 4.43MHz playback colour
(22)	←	5v Supply
(23) 4v0	→	4.43MHz 625kHz playback chroma to bandpass filter to pin 5
(24) 2v2	←	Colour / black and white select
(25) 2v3	←	300mV pk-pk, 625kHz playback chroma
(26) 2v4	←	SP / LP control
(27) 4v6	←	From pin 10 via bandpass filter
(28) 3v5	←	25Hz Drum flip flop pulse

HA11772

Video head playback preamplifier

Package type B (16 pins)

Pins		Function
(1) 0v		
(2)		
(3)	←	Head 1 playback FM
(4) 0v Ground		
(5)	←	Head 2 playback FM
(6)		
(7)		
(8)	←	5v Supply
(9)	→	Playback FM output
(10) n/c		
(11) 0v Ground		
(12) 0v Ground		
(13) n/c		
(14) n/c		
(15)	←	25Hz Head switching flip flop pulse
(16)	←	5v Supply

HA11780

Drum flip flop & vertical pulse generator

Package type B (18 pins)

Pins		Function
(1) 0v Ground		
(2) 0v1	←	Vertical pulse control, H = pin 4 output
(3) 0v Ground		
(4)	→	Vertical pulse for TV synchronization
(5) 0v2		
(6)		
(7)	←	Vertical lock
(8)	←	Head 2 switching control
(9) n/c		
(10)	←	9v Supply
(11) 0v Ground		
(12)	→	25Hz Drum flip flop pulse
(13) 4v9	←	Drum frequency to voltage converted FG pulse
(14)		
(15)	←	Head 1 switching control
(16)		
(17)	←	1v5 pk-pk Drum motor PG pulses
18) 0v Ground		

HA11782

Playback Video head pre amplifier

Package type B (14 pins)

Pins		Function
(1) 0v8		
(2) 2v1	←	Playback from head 1
(3) 0v7		
(4) 0v Ground		
(5) 0v7	←	Playback from head 2
(6) 2v2		
(7) 0v8		
(8)	←	5v Supply
(9)	←	25Hz Drum flip flop pulse
(10) 0v Ground		
(11)	→	0v3 pk-pk Playback FM
(12)	←	5v Supply
(13) 2v6		
(14) 0v Ground		

HA11797

Video head record & playback amplifier

Package type B (30 pins)

Pins		Function
(1)	←	Record mode 9v Supply
(2)	←	Record mode 9v Supply
(3)	→	4v0 pk-pk Record FM to video heads
(4) 0v		
(5) 0v Ground		
(6) 1v8		
(7) 0v7	←	200mV pk-pk playback FM from head 1
(8) 0v Ground		
(9) 0v7	←	200mV pk-pk playback FM from head 2
(10) 1v7		
(11)		
(12)		
(13)	←	Playback mode 5v Supply
(14) 0v		
(15)	←	25Hz, 5v pk-pk flip flop pulse head switching
(16) 0v Ground		
(17)		
(18) 5v0		
(19)	→	300mV pk-pk Playback FM output
(20)	←	150mV pk-pk FM from pin 19 via 1 line drop out compensation circuit
(21)	←	5v playback supply
(22) 3v0		
(23) 3v8	←	Playback Luminance FM from pin 28
(24) 0v8		
(25)	←	Record mode chroma input
(26)	←	Record mode Luminance input
(27)	←	Record mode signal mute control
(28)	→	2v0 pk-pk playback FM to pin 23
(29) 0v		
(30) 0v		

HA118017

Video head playback amplifier **Package type C (28 pins)**

Pins		Function
(1)	←	Head switching control
(2)	←	25Hz head switching pulse
(3)	←	Playback 5v supply
(4)		
(5)	←	Trick play control input
(6)		
(7)	←	Playback from head 1
(8) 0v Ground		
(9)	←	Playback from head 2
(10)		
(11)		
(12)		
(13)		
(14)		
(15) 0v Ground		
(16)		
(17)	←	Head input for AGC detection control
(18)	→	Playback FM output
(19)		
(20)	←	Gain control
(21)	←	Sync' pulse
(22)	←	Record 9v supply
(23)	←	Record mode Luminance FM input
(24)	←	Picture mute
(25)	←	Record mode chroma input
(26) 0v Ground		
(27)	→	Record mode FM signal to heads
(28)	→	Record mode FM signal mute = L

HA118019

4 × Video head SP & LP preamplifier

Package type B (30 pins)

Pins		Function
(1)	2v5 playback	
(2)	⟶	SP head 2 record mode
(3)	⟵	SP head 2 playback mode
(4)	0v Ground	
(5)	⟵	SP head 1 playback mode
(6)	⟶	SP head 1 record mode
(7)	2v3 playback	
(8)	2v3 playback	
(9)	⟶	LP head 1 record mode
(10)	⟵	LP head 1 playback mode
(11)	0v Ground	
(12)	⟵	LP head 2 playback mode
(13)	⟶	LP head 2 record mode
(14)	2v2 playback	
(15)	⟵	5v Supply
(16)	1v6 playback	
(17)	⟵	Record / playback control, record mode = 2v5, playback = 0v5
(18)	⟶	Record FM signal
(19)	⟵	Playback mode luminance
(20)	2v1 playback	
(21)	⟵	AGC luminance from pin 19
(22)	0v Ground	
(23)	⟶	Playback luminance & chroma
(24)	⟵	Control, SP = L, LP = H
(25)	⟶	Chroma switching pulse
(26)	⟵	25Hz Head switching pulse
(27)	2v playback	
(28)	⟵	Cue & review control
(29)	1v8 playback	
(30)	⟵	5v Supply

HA118122

Video head switching

Package type B
(30 pins)

Pins		Function
(1) 2v2		
(2) 0v		
(3)	←	SP, head 1 playback FM
(4)	→	SP, record FM
(5)	←	SP, head 2 playback FM
(6)	←	SP switching control
(7) 2v2		
(8) 0v Ground		
(9) 4v2		
(10)	←	SP switching control
(11)	←	LP, head 1 playback FM
(12)	→	LP, record FM
(13)	←	LP, head 2 playback FM
(14)		
(15) 4v2		
(16)	←	5v Supply
(17)	←	5v Supply
(18)	←	Logic control
(19)	←	25Hz flip flop head switching
(20) 2v1	←	Record mode mute control
(21) 2v7	←	Composite sync' pulses
(22)	←	Logic control
(23)	→	Playback FM
(24) 0v Ground		
(25)	←	Buffered FM from pin 23
(26) 1v5		
(27)	→	Playback FM output
(28) 1v6		
(29)	←	Record FM input
(30)	←	Record mode 9v Supply

HA11827

Drum and Capstan motor servo

Package type B (42 pins)

Pins		Function
(1)	⟶	Capstan normal speed = L, Search mode = H
(2)	⟶	3v5 pk-pk squarewave Capstan speed control
(3)	⟶	Capstan Forward mode = L, Reverse mode = H
(4)	⟶	4v5 pk-pk Drum flip flop pulse
(5)	⟶	Standard, Long play or Extended play select
(6)	⟵	3v0 pk-pk Capstan FG pulse
(7)	⟶	Standard, Long play or Extended play select
(8) 4.43Mhz xtal		
(9) 4.43Mhz xtal		
(10) 2v5		
(11)	⟶	3v5 pk-pk squarewave Drum phase control
(12)	⟶	5v0 pk-pk, 12.5Hz control pulse
(13)	⟵	4v0 pk-pk, 30 Hz control pulse from ACE head
(14)	⟵	Drum flip flop delay control, delay & normal = L
(15)	⟵	1v5 pk-pk Drum PG
(16) Head 1 switching control circuit		
(17) Head 2 switching control circuit		
(18) Drum flip flop delay control circuit		
(19) Vertical lock control circuit		
(20) Vertical lock oscillator RC circuit		
(21)		
(22)	⟶	4v5 pk-pk, Still & search modes vertical pulse
(23)	⟵	Vertical pulse on = H, normal = L control
(24) Tracking control preset		
(25)		
(26)	⟵	5v Supply
(27)	⟶	4v5 pk-pk Record mode control
(28) 0v Ground		
(29)	⟵	0v7 pk-pk Vertical sync'
(30)		
(31)	⟶	5v0 pk-pk Capstan FG pulse
(32)	⟵	Mode control, Record = H, Playback = L
(33)		
(34)	⟵	1v2 pk-pk Drum FG pulse
(35) 0v Ground		
(36)		
(37) 3v0, Drum discriminator preset		

(Continued overleaf)

(38)
(39) ⟶ 3v8 pk-pk squarewave Drum speed control
(40) ⟵ Mode control, Record = H, Playback = L
(41) ⟶ Capstan motor phase control
(42) ⟵ Capstan motor start = H

HA11828NT

Drum and Capstan servo

Package type A (42 pins)

Pins			**Function**
(1)	n/c		
(2)	0v Ground		
(3)		←	Tracking control
(4)	4v1	←	2v4 pk-pk ACE head playback control pulse
(5)	4v2		
(6)	300mV pk-pk squarewave		
(7)			
(8)		←	Capstan brake control from system control
(9)	0v		
(10)		←	1v0 pk-pk Capstan motor FG pulses
(11)		←	Trick speeds = Hi
(12)		→	5v0 pk-pk squarewave Capstan phase to pin 19
(13)		←	Capstan control from system control
(14)			
(15)		→	5v0 pk-pk squarewave Capstan speed to pin 20
(16)	0v		
(17)		←	Trick speeds = Hi
(18)	3v0	→	Capstan motor drive
(19)	2v5	←	From pin 12
(20)	2v5	←	From pin 15
(21)	3v0		
(22)	2v3	←	400mV pk-pk, 4.43Mhz clock pulse
(23)	2v3		
(24)	2v6	→	Drum motor drive
(25)	2v2	←	Drum motor speed control from pin 27
(26)	2v3	←	Drum motor phase control from pin 28
(27)		→	To pin 25
(28)		→	To pin 26
(29)		←	Reverse control
(30)	2v5		
(31)		→	5v0 pk-pk squarewave 25Hz switching pulse
(32)			
(33)		←	Vertical jitter manual adjustment
(34)			
(35)	2v5 pk-pk Vertical sync' pulse		
(36)		→	Vertical sync' pulse
(37)		←	2v5 pk-pk Drum motor FG pulses

(Continued overleaf)

HA11828NT

(38)
(39) ⟶ 5v0 pk-pk squarewave record mode to ACE head
(40) ⟵ 5v Supply
(41) ⟵ 2v0 pk-pk composite sync' pulses
(42) n/c

HA11839NT

Video head playback and record amplifier

Package type B (30 pins)

Pins		Function
(1) 0v Ground		
(2)	←	9v2 Record mode
(3) 4v1	→	3v5 pk-pk Record FM to heads
(4) 0v3		
(5) 0v Ground		
(6)		
(7)		
(8) 0v Ground		
(9)	←	Head 1 playback FM
(10) 1v7		
(11)	←	Head 2 playback FM
(12) 1v6		
(13)	←	Playback mode 5v supply
(14)	←	25Hz Drum flip flop head switching pulse
(15)		
(16) 0v Ground		
(17) 0v Ground		
(18)	←	Playback 5v supply
(19)	→	300mV pk-pk playback FM
(20)		
(21)	←	Playback 5v supply
(22)		
(23) 3v6	←	Playback FM automatic gain control
(24) 0v7		
(25)	←	Record mode chroma
(26)	←	Record mode FM luminance
(27)	←	Record mode = L, Playback mode = H
(28)	→	400mV pk-pk playback FM to chroma circuit
(29) 2v9		
(30)		

HA11843

Servo processor **Package type B (42 pins)**

Pins		Function
(1) n/c		
(2) 0v Ground		
(3)	←	2v5 pk-pk ramp waveform & Tracking control
(4)	←	2v5 pk-pk Capstan ACE head control pulses
(5) 4v2	→	SP or LP detection
(6) 0v3 pk-pk squarewave		
(7)		
(8)		
(9)		
(10)	←	1v0 pk-pk Capstan FG pulses
(11)	←	Search = H
(12)	→	5v0 pk-pk squarewave, playback & record Capstan phase to pin 19
(13)		
(14)		
(15)	→	4v5 pk-pk squarewave, playback & record Capstan speed to pin 20
(16)		
(17)	←	Pause & frame advance mode control
(18)	→	Capstan speed output
(19)	←	From pin 12
(20)	←	From pin 15
(21) 3v0		
(22)	←	2v5 pk-pk, 4.43Mhz clock pulse
(23) 2v3		
(24)	→	Drum drive
(25)	←	Drum speed control from pin 27
(26)	←	Drum phase control from pin 28
(27)	→	4v0 pk-pk Drum speed control to pin 25
(28)	→	to pin 26
(29)	←	Reverse search control
(30)		
(31)	→	5v0 pk-pk, squarewave vertical pulse
(32)		
(33)	←	Vertical jitter manual control
(34)		
(35)	←	Head switching point manual adjustment

(Continued opposite)

(36)	⟶	Vertical pulse out
(37)	⟵	2v5 pk-pk Drum FG pulses
(38)		
(39)	⟶	5v0 pk-pk squarewave, record mode to ACE control head
(40)	⟵	5v Supply
(41)	⟵	5v0 pk-pk Vertical sync' input
(42) n/c		

HA11870NT

Head pre-amplifier **Package type B (30 pins)**

Pins		Function
(1)		
(2)	⟶	Record mode, SP Head 2
(3)	⟵	Playback mode, SP Head 2
(4) 0v Ground		
(5)	⟵	Playback mode, SP Head 1
(6)	⟶	Record mode, SP Head 1
(7)		
(8)		
(9)	⟶	Record mode, LP Head 1
(10)	⟶	Playback mode, LP Head 1
(11) 0v Ground		
(12)	⟵	Playback mode, LP Head 2
(13)	⟶	Record mode, LP Head 2
(14)		
(15)	⟵	5v Supply
(16)		
(17)	⟵	Record start mute = Hi
(18)	⟵	Record mode FM
(19)	⟶	Playback FM detection
(20)		
(21)	⟵	From pin 23 via High pass filter for AGC circuit
(22) 0v Ground		
(23)	⟶	Playback FM output
(24)	⟵	SP / LP control
(25) n/c		
(26)	⟵	25Hz Drum Flip flop pulse
(27) 0v Ground		
(28)	⟵	5v Supply
(29)	⟵	5v Supply
(30)	⟵	5v Supply

Note :
Pins 8,9,10,12,13 &14 not connected if no LP facility
Pin 24 grounded if no LP facility

HA13008

Drum motor controller

Package type B (20 pins)

Pins		Function
(1)	←	Phase & speed control from servo
(2)		
(3)	→	Bias to Hall effect device 1
(4)	←	Switching bias from Hall effect device 1, stator coil selection
(5)	←	Switching bias from Hall effect device 1, stator coil selection
(6)		
(7)	→	Stator coil 1 control current
(8)	←	Stator coil 1 control current
(9)		
(10) Stator coils common return resistor		
(11)		
(12)		
(13)	→	Stator coil 2 control current
(14)	←	Stator coil 2 control current
(15)	→	Bias to Hall effect device 3
(16)	←	Switching bias from Hall effect device 2, stator coil selection
(17)	←	Switching bias from Hall effect device 2, stator coil selection
(18)	→	Bias to Hall effect device 2
(19)		
(20)	←	Drum motor start

HA13403

Drum motor drive

Package type B (23 pins)

Pins		Function
(1)	←	12v Supply
(2) 12v		
(3) 0v Ground		
(4) 4v7		
(5) 2v7	←	Servo drive
(6) 2v5		
(7)	←	5v Supply
(8) 4v6		
(9)	←	+ve FG pulses from motor
(10) 2v8	→	Motor speed FG pulses to servo circuit
(11)	←	–ve FG pulses from motor
(12) 0v Ground		
(13) 2v1	←	+ve pulses Hall effect device 1
(14)	←	Output current sensing feedback from pin 16
(15) 2v1	←	–ve pulse Hall effect device 1
(16)	→	To pin 14
(17) 2v1	←	+ve pulses Hall effect device 2
(18) 8v8	→	Stator coil 3 drive
(19) 2v1	←	–ve pulses Hall effect device 2
(20) 8v8	→	Stator coil 2 drive
(21) 2v1	←	+ve pulse Hall effect device 3
(22) 8v8	→	Stator coil 3 drive
(23) 2v1	←	–ve pulse Hall effect device 3

HA13430A

Drum motor drive **Package type B (16 pins)**

Pins		Function
(1) 2v7	←	Hall effect device pulses
(2) 2v6	←	Hall effect device pulses
(3) 2v6	←	Drive from servo
(4) 0v Ground		
(5) 0v Ground		
(6) 4v4		
(7)	→	6v0 pk-pk drive to stator coil 1
(8) 0v current sensing resistor		
(9)	→	6v0 pk-pk drive to stator coil 2
(10)	←	5v Supply
(11) 1v pk-pk, 10mSec' squarewave		
(12) 0v Ground		
(13) 0v Ground		
(14)	→	2v0 pk-pk Amplified Drum FG pulses
(15)	←	Drum motor FG –ve pulses
(16)	←	Drum motor FG +ve pulses

HA13457NT

Capstan motor drive

Package type B (24 pins)

Pins		Function
(1)	←	5v Supply
(2)	←	Servo drive control signal
(3)		
(4)	→	FG pulse output
(5)	←	FG coil pulses
(6)	←	FG coil pulses
(7)	←	Forward & reverse control
(8)		
(9)		
(10)		
(11) 0V Ground		
(12)		
(13)		
(14)		
(15)	→	Stator coil drive 1
(16)	→	Stator coil drive 2
(17)	→	Stator coil drive 3
(18)	←	15v Supply
(19)	←	+ve Hall effect device 1
(20)	←	–ve Hall effect device 1
(21)	←	+ve Hall effect device 2
(22)	←	–ve Hall effect device 2
(23)	←	+ve Hall effect device 3
(24)	←	–ve Hall effect device 3

HD49716

Capstan & drum servo

Package type B (64 pins)

Pins			Function
(1)	0v	←	search speed control
(2)	0v3	←	Reverse & brake control
(3)			
(4)		←	5v Supply
(5)	n/c		
(6)	0v2	←	Pause & slow control
(7)	0v2	←	Drum on = L
(8)		←	300mV pk-pk, 4.43Mhz clock pulse
(9)		←	Speed play control
(10)	0v, 5v0 pk-pk squarewave (control pulse)		
(11)		←	Record mode 5v0 Supply
(12)		←	Record & playback capstan FG pulses
(13)			
(14)	0v2	←	Capstan slow / pause control
(15)		←	Capstan search mode control
(16)	1v8	→	Record mode 600mV pk-pk pulses to –ve ACE control head
(17)	0v	→	Record mode 5v0 pk-pk squarewave to +ve ACE control head
(18)		←	5v Supply
(19)	3v0	→	Capstan motor drive
(20)	2v5		
(21)	2v8		
(22)	2v5		
(23)	0v	←	Capstan phase signal from pin 24
(24)	2v5	→	To pin 23
(25)		←	5v Supply
(26)	2v5		
(27)	2v8	←	2v5 pk-pk squarewave amplified and shaped playback control pulses from ACE head
(28)	2v8		
(29)	3v4		
(30)	2v8	→	3v0 pk-pk playback ACE control pulses to system control circuit 5v0 pk-pk Record ACE control pulses to system control
(31)	0v		
(32)		←	–ve Drum tacho pulses

(Continued overleaf)

HD49716

(33) ←—— 400mV pk-pk +ve Drum tacho pulses
(34) 2v0
(35) 2v7
(36) 2v5 ——→ 500mV pk-pk record & playback drum phase signal to pin 42
(37) 0v
(38) 2v5
(39) 2v8
(40) 2v8
(41) 0v
(42) 2v8 ←—— 300mV pk-pk Drum phase from pin 36
(43) 2v7
(44) 2v8
(45) 2v8 ——→ Drum motor drive
(46) 0v Ground
(47) 0v Ground
(48) ——→ 25Hz flip flop switching pulse
(49) ←—— Record index data
(50) 0v Ground
(51) 0v Ground
(52) Record & playback 5v0 pk-pk Horizontal sync' pulses
(53) n/c
(54) n/c
(55)
(56) ←—— Head switching adjustment
(57)
(58) 1v8 ←—— 2v0 pk-pk Composite sync' pulse
(59) 5v0
(60) ←—— Tracking control
(61) ←—— 4v0 pk-pk Drum FG pulses
(62) ——→ Composite sync' noise detection
(63) ←—— Record on / off control
(64) ←—— SP speed control

HD49722NT

Drum & Capstan motor servo

Package type B (64 pins)

Pins			Function
(1)	Picture search (double normal speed) = L		
(2)		←	LP mode
(3)		←	SP mode = L
(4)		←	SP = H
(5)			
(6)	0v Ground		
(7)	0v2		
(8)			
(9)		←	Vertical picture lock control
(10)	0v2	←	Playback head switching point control
(11)		←	4v0 pk-pk Drum PG & FG pulses
(12)	2v7		
(13)	5v0		
(14)	n/c		
(15)	2v6	→	Drum motor drive
(16)	2v5	←	From pin 17
(17)	2v5	→	Frequency to voltage converted Drum PG pulses to pin 16
(18)	2v4		
(19)		←	From pin 20
(20)	2v4	→	Drum phase control to pin 19
(21)	2v6		
(22)		→	Capstan phase control to pin 23
(23)		←	From pin 22
(24)		→	Converted frequency to voltage capstan FG pulses to pin 27
(25)	2v6		
(26)	2v6		
(27)		←	From pin 24
(28)			
(29)	0v4	→	Capstan motor drive
(30)		←	5v Supply
(31)			
(32)			
(33)		→	Record mode 5v pk-pk control tracking pulses to ACE head
(34)		←	Playback mode control tracking pulses from ACE head

(Continued overleaf)

HD49722NT

(35) 2v6	⟶	Playback ACE control tracking pulses to pin 38
(36) 2v6		
(37) 0v Ground		
(38)	⟵	From pin 35
(39)	⟶	Amplified ACE control track pulses to internal pulse square pin 40
(40)	⟵	From pin 39
(41)	⟵	FG pulse from pin 42
(42)	⟶	2v4 pk-pk Capstan FG pulse to pin 41
(43)	⟵	FG pulse from capstan motor
(44)	⟵	5v Supply
(45) 0v		
(46)	⟵	Index mode control
(47)		
(48) 0v		
(49)	⟵	NTSC = H, PAL & SECAM = L
(50)	⟵	Serial Data
(51)	⟵	Serial clock pulse
(52)	⟶	Capstan FG pulse to system control
(53)	⟶	Playback control tracking pulse to system control
(54) 4v6 pk-pk tracking pulse		
(55)	⟵	Composite video sync' pulses
(56) 0v		
(57) n/c		
(58)	⟵	System clock frequency × 3 for internal capstan & Drum servo reference from video circuit
(59)	⟶	Inverted system clock frequency × 3
(60)		
(61)		
(62)	⟶	5v0 pk-pk Drum flip flop pulse to video head pre amplifier
(63)		Luminance noise reducing pulse to video head pre amplifier
(64)	⟶	Vertical pulse

Pins 45, 48, 56 may be grounded

HD49733

Drum & Capstan Servo

Package type B (56 pins)

Pins		Function
(1) n/c		
(2)	→	Drum motor start
(3)	→	SP = L
(4)	→	FM Video digital tracking control
(5) 0v Ground		
(6)	←	Video vertical sync' pulse
(7)	←	Playback head switching point control
(8) 0v Ground		
(9)	←	Drum FG & PG pulses
(10) n/c		
(11)	←	Switched 5v Supply
(12)	→	Drum motor control
(13) Drum gain feedback circuit		
(14)	→	Drum control converted frequency to voltage to feed-back circuit
(15)		
(16)		
(17)		
(18)		
(19)	←	Capstan & Drum phase detection
(20)	→	Capstan phase control to pin 21 via loop filter
(21)	←	Capstan phase control
(22)	→	Capstan, converted frequency to voltage signal to pins 24, 25 & 26 via capstan drive gain switching circuit
(23)	→	Capstan, converted frequency to voltage signal to pins 24, 25 & 26 via capstan drive gain switching circuit
(24)	←	From pins 21 & 22
(25)	←	From pins 21 & 22
(26)	←	From pins 21 & 22
(27)	→	Capstan motor drive output
(28)	←	5v Supply
(29)	→	Record mode control track pulses to ACE head
(30)	←	Playback mode control track pulses from ACE head
(31) 0v Ground		
(32) 2v6		
(33)		
(34) Control track pulse amplifier gain circuit		

(Continued overleaf)

HD49733

(35) ⟶ Amplified Playback control track pulses to pin 36
(36) ⟵ From pin 35
(37) ⟵ Capstan FG pulses
(38) ⟵ 5v Supply
(39) ⟶ Control clock pulse to system control i.c
(40) ⟷ Record & playback data from system control i.c
(41) 0v Ground
(42) ⟵ 10 MHz clock pulse
(43)
(44)
(45)
(46) ⟶ Drum flip flop pulse
(47) ⟶ Luminance noise detection pulse
(48) ⟶ Vertical pulse
(49) Vertical pulse on = L
(50) ⟵ Composite sync' pulse
(51) 0v Ground
(52) n/c
(53) ⟵ Serial Data
(54) ⟵ Serial clock
(55) ⟶ Capstan FG pulse to system control
(56) ⟶ Control track pulse to system control

HD49756

Drum & Capstan Servo

Package type B (56 pins)

Pins		Function
(1) 3v6		
(2)	⟶	LP mode = H
(3) 0v		
(4) 5v0		
(5) 0v Ground		
(6) 0v2	⟵	Vertical lock manual adjustment
(7) 2v1	⟵	Drum PG pulse
(8) 0v Ground		
(9) 2v7	⟵	Drum FG pulse
(10) 2v7		
(11) 2v7		
(12) 1v4	⟶	Drum motor control output
(13) 2v5	⟵	Converted Drum FG frequency to voltage from pin 14
(14) 2v5	⟶	Converted Drum FG frequency to voltage to pin 13
(15) 2v5	⟵	Drum phase control from pin 17
(16) 2v5		
(17) 2v5	⟶	Drum phase control to pin 15
(18) 2v3		
(19) NTSC = H, PAL = L		
(20) 2v5	⟶	Capstan phase control to pin 21
(21) 2v5	⟵	Capstan phase control from pin 20
(22) 2v5	⟶	Converted Capstan FG frequency to voltage to pin 24
(23) 2v5		
(24) 2v5	⟵	Converted Capstan FG frequency to voltage from pin 22
(25) 2v5		
(26) 2v6		
(27) 2v6	⟶	Capstan motor control
(28)	⟵	5v Supply
(29)	⟶	Record mode control pulses to ACE head
(30)	⟵	Playback control pulses from ACE head
(31) 0v Ground		
(32) 2v5		
(33) 2v5		
(34) 2v5		
(35) 2v5	⟶	Amplified playback ACE control pulse to pin 36
(36) 2v5	⟵	Control pulse from pin 35

(Continued overleaf)

HD49756

(37) ←—— Capstan FG pulse
(38) ←—— 5v Supply
(39) ——→ Clock control to system control circuit
(40) ←——→ System micro' control
(41) 0v
(42) 2v6 ←—— Composite sync' pulses
(43) 2v6
(44)
(45) ——→ 25Hz Drum flip flop pulse
(46)
(47) 2v5
(48) ←—— Trick modes vertical lock sync' pulse
(49) 0v
(50) ←—— Composite sync' pulses
(51) ←—— 5v Supply
(52) 3v8 Power on reset
(53) ←—— Serial Data
(54) ←—— Serial clock pulse
(55) ——→ Capstan FG pulse to system control circuit
(56) ——→ Control pulse to system control circuit

2 × 2 pole, 4 way switches

Package type B
(15 pins)

Pins		Function
(1)	⟶	Switch 1 terminal 1
(2)	⟶	Switch 1 terminal 3
(3)	⟵	Switch 1 pole
(4)	⟶	Switch 1 terminal 4
(5)	⟶	Switch 1 terminal 2
(6)	⟵	Logic input selects poles 1 & 2 position
(7)		
(8)	0v Ground	
(9)	⟵	Logic input selects poles 1 & 2 position
(10)	⟵	Logic input selects poles 1 & 2 position
(11)	⟶	Switch 2 terminal 4
(12)	⟶	Switch 2 terminal 1
(13)	⟵	Switch 2 pole
(14)	⟶	Switch 2 terminal 2
(15)	⟶	Switch 2 terminal 3
(16)	⟵	12v Supply

Switch designated numbers, i.e 1,2, are only given to distinguish between the individual devices
For clarity, the signal paths assume input via the pole terminal and exit via one of the other terminals. In practice the signal flow may be in any direction.

HES8455

Chroma processor

Package type B (22 pins)

Pins			Function
(1)	3v5	⟶	600mV pk-pk, 627kHz playback chroma output
(2)	0v	⟵	4.43MHz playback chroma from pin 3
(3)	0v	⟶	500mV pk-pk chroma to pin 2
(4)	2v8	⟵	1v0 pk-pk record mode video
(5)		⟵	5v Supply
(6)		⟶	600mV pk-pk , 627kHz record mode chroma output
(7)		⟵	Playback 627kHz chroma input
(8)	0v Ground		
(9)		⟵	Trick speed control
(10)		⟵	LP mode control
(11)		⟵	Automatic phase control
(12)	2v8	B&W / colour select	
(13)	1v7	⟶	4.43MHz, 800mV pk-pk oscillator output
(14)	n/c		
(15)		⟵	5v0 pk-pk 25Hz flip flop pulse
(16)		⟵	5v0 pk-pk Vertical sync' pulse
(17)		⟵	Secam select = H
(18)		⟶	1v0 pk-pk, 4.43MHz playback chroma to pin 19 via 2 line delay
(19)		⟵	200mV pk-pk from delay line
(20)		⟵	5v Supply
(21)	n/c		
(22)		⟵	SP mode = H

IR3702

4 × Operational amplifiers

Package type B (14 pins)

Pins		Function
(1)	→	Amplifier 1 output
(2)	←	Amplifier 1 inverted input
(3)	←	Amplifier 1 non inverted input
(4)	←	18v Supply
(5)	←	Amplifier 2 non inverted input
(6)	←	Amplifier 2 inverted input
(7)	→	Amplifier 2 output
(8)	→	Amplifier 3 output
(9)	←	Amplifier 3 inverted input
(10)	←	Amplifier 3 non inverted input
(11) 0v Ground		
(12)	←	Amplifier 4 non inverted input
(13)	←	Amplifier 4 inverted input
(14)	→	Amplifier 4 output

Amplifiers designated numbers, i.e 1,2, etc are given to distinguish between the individual devices

L2722

Dual operational amplifiers

Package type B (8 pins)

Pins		Function
(1)	⟶	Amplifier 1 output
(2)	⟵	14v Supply
(3)	⟶	Amplifier 2 output
(4)	0v Ground	
(5)	⟵	Amplifier 2 inverted input
(6)	⟵	Amplifier non inverted input
(7)	⟵	Amplifier 1 non inverted input
(8)	⟵	Amplifier 1 inverted input

Example : used as a loading motor driver
Stop mode Pin 1, 3, 6, 7 = 0v
Forward mode Pin 1 = 0v, pin 3 = 14v, pin 6 = 5v, pin 7 = 0v
Reverse mode pin 1 = 14v, pin 3 = 0v, pin 6 = 0v, pin 7 = 5v

LA1363

Sound FM detector

Package type B (14 pins)

Pins		**Function**
(1) 1v8		
(2) 1v8	←	Sound I.F input
(3) 0v Ground		
(4) 0v Ground		
(5)	←	10v Supply
(6) 1v7	←	Sound mute control
(7) 6v5		
(8) 5v5	→	Audio output
(9) FM detector LC circuit		
(10) FM detector LC circuit		
(11) n/c		
(12)	→	Separate audio amplifier 1 output
(13)	←	Separate audio amplifier 1 input
(14)	←	Separate audio amplifier 1 input

Pins 12,13 & 14 are connected to an independent amplifier within the device. When not used pins 13 & 14 may be connected to 0v Ground.

LA1365

Sound FM detector

Package type B (14 pins)

Pins		**Function**
(1) 1v5		
(2) 1v8	←	Sound I.F input
(3) 0v Ground		
(4) 0v Ground		
(5)	←	10v Supply
(6) 1v5	←	Sound mute control
(7) 6v5		
(8) 5v3	→	Audio output
(9) FM detector LC circuit		
(10) FM detector LC circuit		
(11) n/c		
(12)	→	Separate audio amplifier 1 output
(13)	←	Separate audio amplifier 1 input
(14)	←	Separate audio amplifier 1 input

Pins 12,13 & 14 are connected to an independent amplifier within the device. When not used pins 13 & 14 may be connected to 0v Ground.

2 × single pole, 2 way switches

Package type A (9 pins)

Pins		Function
(1)	←	Switch 1, input terminal A
(2)	←	Switch 1, input terminal B
(3)	→	Switch 1 pole terminal output
(4)	←	Switches input selection control logic
(5) 0v Ground		
(6)	←	8v Supply
(7)	→	Switch 2 pole terminal output
(8)	←	Switch 2, input terminal A
(9)	←	Switch 2, input terminal B

Switch 2 signal direction is shown leaving from the pole terminal. In practice it may be in any direction. Switch 1 however, has a buffer amplifier between input pin 1 or 2 and the output pin 3.

LA5611

Power supply regulator/sampler

Package type B (14 pins)

Pins		Function
(1)	⟶	Always 5v
(2)	⟵	Power on / off control
(3)		
(4)	⟶	Switched 5v
(5)	⟵	6v Supply
(6)		
(7)		
(8)		
(9)	⟶	on/off switching for external use
(10)		
(11)		
(12)	⟵	Control sensing input
(13)	⟵	14v Supply
(14)		

LA6324N

4 × Operational amplifiers

Package type B (14 pins)

Pins		Function
(1)	→	Amplifier 2 output
(2)	←	Amplifier 2 Inverting input
(3)	←	Amplifier 2 non inverting input
(4)	←	15v Supply (typical)
(5)	←	Amplifier 3 non inverting input
(6)	←	Amplifier 3 inverting input
(7)	→	Amplifier 3 output
(8)	→	Amplifier 1 output
(9)	←	Amplifier 1 inverting input
(10)	←	Amplifier 1 non inverting input
(11) 0v Ground		
(12)	←	Amplifier 4 non inverting input
(13)	←	Amplifier 4 inverting input
(14)	→	Amplifier 4 output

Amplifiers designated numbers, i.e 1,2, etc are given to distinguish between the individual devices

LA6358

Dual operational amplifier

Package type B (8 pins)

Pins		Function
(1)	——→	Amplifier 1 output
(2)	←——	Amplifier 1 inverted input
(3)	←——	Amplifier 1 non inverted input
(4) 0v Ground		
(5)	←——	Amplifier non inverted input
(6)	←——	Amplifier 2 inverted input
(7)	——→	Amplifier 2 output
(8)	←——	30v Supply

LA6393D

Dual comparator

Package type B
(8 pins)

Pins		Function
(1)	→	Comparator 1 output
(2)	←	Comparator 1 inverted input
(3)	←	Comparator 1 non inverted input
(4)	0v Ground	
(5)	←	Comparator 2 non inverted input
(6)	←	Comparator 2 inverted input
(7)	→	Comparator 2. output
(8)	←	12v Supply

LA6393S

Dual comparator

Package type A (9 pins)

Pins		Function
(1)	←	12v Supply
(2)	→	Comparator 1 output
(3)	←	Comparator 1 inverted input
(4)	←	Comparator 1 non inverted input
(5)	0v Ground	
(6)	←	Comparator 2 non inverted input
(7)	←	Comparator 2 inverted input
(8)	→	Comparator 2 output
(9)	←	12v Supply

LA7016

Single pole 2 way switch

Package type A (8 pins)

Pins		Function
(1)	←	8v supply
(2)	→	Terminal 1output
(3)	←	Switch control voltage (H = Terminal 1, or L = Terminal 2)
(4)	←	Pole connection input
(5)	0v Ground	
(6)		
(7)	→	Terminal 2 output
(8)		

For clarity, above assumes the signal enters at the pole connection and outputs at either terminal 1 or 2 depending on the control switch voltage. In practice, the signal may be in any direction.

LA7031

Record & playback Luminance processor

Package type B (30 pins)

Pins		Function
(1) 5v5		
(2) 2v0	←	Composite video
(3) 8v0		
(4) 0v8	→	Separated sync' pulses
(5) 6v0		
(6) 5v9	→	2v5 pk-pk playback video
(7)	←	Trick modes vertical pulse
(8) 1v2	←	150mV pk-pk Playback chroma
(9) 0v Ground		
(10) 4v3		
(11) 4v3		
(12) 4v2		
(13) 5v0	←	Playback luminance from FM demodulator pin 25
(14) 4v4	←	400mV pk-pk record mode luminance from pin 25
(15) 5v0	→	1v0 pk-pk Luminance, record & playback modes
(16) 4v4		
(17) 3v8	←	100mV pk-pk head playback luminance
(18) 0v Ground		
(19) 4v6, FM limiter circuit		
(20) 4v6, FM limiter circuit		
(21)	←	9v Supply
(22) 5v1, FM demodulator circuit		
(23) 5v2, FM demodulator circuit		
(24) 1v9		
(25) 7v6	→	Playback Luminance to pin 13 & Record luminance to pin 14
(26) 0v		
(27)	←	External & Tuner composite Video
(28)		
(29)	←	1v pk-pk playback luminance from pin 25
	→	1v pk-pk luminance record mode to head amplifier circuit
(30) 4v7		

Video head playback & record amplifier

Package type B (30 pins)

Pins		Function
(1) 0v1	←	Record mode luminance from pin 29
(2) 0v		
(3) 0v	←	50mV pk-pk, 627kHz Record mode chroma
(4) 0v	→	500mV pk-pk, Record mode chroma to video heads
(5) 4v0		
(6) 0v5	←	Playback FM from head 1
(7) 0v Ground		
(8) 0v6	←	Playback FM from head 2
(9) 4v0		
(10) 7v8		
(11) 0v, FM modulator CR circuit		
(12) 0v, FM modulator CR circuit		
(13) 0v		
(14)	←	Record mode 9v Supply
(15)	←	Dark clip adjustment
(16)	←	White clip adjustment
(17) 0v		
(18) 0v		
(19) 0v4	←	400mV pk-pk, record mode E-E luminance
(20) 0v		
(21) 2v5	←	25 Hz Head switching pulse
(22) 0v Ground		
(23) 6v3	→	500mV pk-pk playback FM luminance output
(24) 2v8		
(25) 3v5		
(26) 0v7		
(27)	←	9v5 Supply
(28) 5v0	←	Playback luminance from pin 29
(29) 4v8	→	Record & playback luminance
(30) 0v		

LA7035

Colour processor

Package type A (30 pins)

Pins		Function
(1) 0v	⟶	Record mode colour 627kHz
(2) 6v4	⟶	2v5 pk-pk Playback chroma to delay line
(3) 1v0		
(4) 5v3	⟵	Record & playback chroma from 4.43MHz bandpass filter
(5) 1v3		
(6) 0v	⟵	Vertical sync' pulses
(7) 0v		
(8) 0v Ground		
(9) 0v7		
(10) 5v3		
(11) 5v4		
(12) 5v2		
(13) 7v0, 4.43MHz xtal		
(14) 3v0, 400mV pk-pk 4.43MHz xtal		
(15) 3v0		
(16) 3v2		
(17) 3v1	⟵	1v5 pk-pk, 4.43MHz Record & E-E mode
(18) 5v5		
(19) 7v0	⟶	800mV pk-pk, 5.06MHz to bandpass filter
(20) 7v2		
(21) 7v2	⟶	800mV pk-pk, Playback chroma output
(22)	⟵	9v Supply
(23) 4v4	⟶	800mV pk-pk Playback chroma to pin 4
(24) 0v		
(25) 5v0		
(26) 3v5	⟵	627kHz playback chroma from head circuit
(27) 6v0		
(28) 6v3	⟵	200mV pk-pk, 5.06MHz from pin 19
(29) 4v8		
(30) 4v6	⟵	Delayed chroma from pin 2

LA7042

Audio Record & playback

Package type B (20 pins)

Pins		Function
(1)		
(2)	←	PB from ACE head
(3)	←	PB level manual control
(4)	→	Amplified Playback to pins 8 & 9
(5)	←	Record / playback mode control
(6) 0v Ground		
(7)		
(8)	←	From pin 4 to internal AGC circuit
(9)	←	From pin 4
(10)	→	Audio to line out socket
(11)		
(12)		
(13)		
(14)	→	Record mode, audio from pin 10 via record manual control
(15)	→	Record mode, audio to record head
(16)	←	Audio mute control
(17)	←	Playback = H, Record & E -E = L
(18)	←	9v Supply
(20)	←	External or TV audio input

LA7050

RF Convertor

Package type A (9 pins)

Pins		Function
(1)	→	Modulated video output
(2)	←	Composite Video input
(3)	→	Modulated sound output
(4)	←	Audio input
(5) 0v Ground		
(6) 9v0		
(7)	←	9v Supply
(8)	←	9v Supply
(9) 0v Ground		

LA7054

RF convertor

Package type A (9 pins)

Pins		Function
(1)	←	Composite video input
(2)	←	Test signal oscillator on / off switch
(3)	←	5v Supply
(4)	←	Audio input
(5)	0v Ground	
(6)	FM modulator LC circuit	
(7)	→	Audio to modulator
(8)	→	Regulated supply to modulator bias
(9)	→	Video to modulator

LA7090

Audio record & playback processor

Package type B (22 pins)

Pins		Function
(1) 7v2		
(2) 7v9		
(3) 0v Ground		
(4) 0v6	←	Playback from audio head
(5) 0v Ground		
(6) 0v Ground		
(7) 0v		
(8) 3v0	→	Amplified playback audio to pin 10
(9) 7v0		
(10)	←	From pin 8
(11)	←	Record mode audio input
(12) 0v6		
(13) 3v4	→	Playback audio output or record mode audio to pin 15
(14) 0v4		
(15) 1v5	←	Record mode audio from pin 13
(16) 0v Ground		
(17)		
(18) 3v6	→	Record mode audio to ACE head
(19) 0v Ground		
(20)	←	Audio mute control on = H
(21)	←	Record = 0v, Playback = 5v0
(22)	←	12v Supply

LA7116

Servo pulse shaper & amplifier

Package type B (24 pins)

Pins		Function
(1) 2v8	——→	4v0 pk-pk squarewave amplified control pulse to servo
(2) 2v5	←——	4v0 pk-pk squarewave amplified control pulse from pin 3
(3) 2v5	——→	To pin 2
(4) 2v5		
(5) 2v5	←——	Control pulse from ACE head
(6) 0v Ground		
(7) 2v6	——→	Drive to drum motor circuit
(8) 2v6		
(9) 2v6	←——	Drum speed control from servo
(10) 2v7	←——	Capstan speed control from servo
(11) 2v7		
(12) 2v8	——→	Drive to capstan motor circuit
(13)	←——	5v Supply
(14) 2v5	——→	4v0 pk-pk drum tacho pulse to servo
(15) 2v5	←——	Tacho pulses from drum motor
(16) 2v4	←——	Drum FG pulses
(17) 2v5		
(18)	——→	4v0 pk-pk drum FG pulses to servo
(19)	←——	5v Supply
(20) 2v5		
(21) 2v5	←——	400mV pk-pk Capstan FG pulses
(22) 2v5		
(23) 2v5	——→	4v0 pk-pk Capstan FG pulses to servo
(24) 5v0		

LA7122S

Drum & Capstan servo

Package type B (30 pins)

Pins		Function
(1)	0v Ground	
(2)		
(3)	40mV / 40msec sawtooth waveform	
(4)	——→	1v1 pk-pk control pulse (Test point)
(5)		
(6)		
(7)	——→	3v3 pk-pk control pulse to ACE head
(8)	——→	3v3 pk-pk control pulse to ACE head
(9)	0v Ground	
(10)		
(11)	←——	Remark input
(12)	←——	5v0 pk-pk record pulse
(13)	←——	Still, frame advance & slow mode control (or ground)
(14)	←——	Switched 5v supply
(15)	——→	3v7 pk-pk control pulse
(16)	——→	Drum motor drive
(17)	←——	Drum motor phase control
(18)	←——	Drum motor speed control
(19)	←——	Capstan motor speed control
(20)	←——	Capstan motor phase control
(21)	——→	Capstan motor drive
(22)	←——	Switched 5v supply
(23)		
(24)		
(25)	←——	22mV pk-pk capstan motor FG pulse
(26)	——→	Capstan motor amplified FG pulse
(27)	——→	4v0 pk-pk drum motor amplified FG pulse
(28)	←——	12mV drum motor FG pulse
(29)	←——	150mV pk-pk drum motor PG pulse
(30)	——→	4v0 pk-pk drum motor amplified PG pulse

LA7214

Sync' pulse separator

Package type A (9 pins)

Pins		**Function**
(1)	⟶	Horizontal sync' pulse to pin 6
(2)	⟵	Video from buffer output pin 4
(3)	⟵	Composite Video input
(4)	⟶	To pin 2
(5) 0v Ground		
(6)	⟵	Horizontal sync' from pin 1
(7)		
(8)	⟵	Vertical sync' pulse ouput
(9)	⟶	5v Supply

LA7221

Video source select

Package type A (9 pins)

Pins		Function
(1)	←	Video from Tuner / I.F
(2)	←	Input source select control 1
(3)	←	AV2 from scart or other external source
(4)	←	Input source select control 2
(5)	0v Ground	
(6)	←	AV1 input from scart or other external source
(7)	←	Playback output mute = 5v
(8)	→	Selected video output

LA7222

2 × single pole, 2 way audio/video signal buffered output switches

Package type A (12 pins)

Pins		**Function**
(1) n/c		
(2)	←	Switch 1 position control logic
(3) 0v Ground		
(4)	←	Switch 2 position control logic
(5)	←	Switch 2, input terminal A
(6) n/c		
(7)	←	Switch 2, input terminal B
(8)	→	Switch 2, buffered output
(9)	←	12v Supply
(10)	←	Switch 1, input terminal B
(11)	→	Switch 1, buffered output
(12)	←	Switch 1, input terminal A

Switch designated numbers, i.e 1,2, are only given to distinguish between the individual devices.

LA7224

Infra red receiver

Package type A
(8 pins)

Pins		Function
(1) 5v1	→	Data to micro’
(2) 5v0		
(3) 5v2		
(4) 3v7		
(5) 0v Ground		
(6) 0v8		
(7) 1v4	←	Infra red receiver diode
(8)	←	5v Supply

LA7225

Infra red receiver

Package type A
(9 pins)

Pins		Function
(1) 6v4	→	Data to micro'
(2) 6v2		
(3)		
(4) 4v7		
(5) 0v Ground		
(6) 0v7		
(7)	←	Data from infra red diode
(8)	←	6v0 Supply
(9) 0v		

LA7226

Infra red receiver | **Package type A (9 pins)**

Pins		**Function**
(1)	⟶	Data to micro'
(2)		
(3) CR Peak detector tuned circuit		
(4)		
(5) 0v Ground		
(6)		
(7)	⟵	Data from infra red diode
(8)	⟵	12v Supply
(9)	⟵	TV remote facility enable

LA7282

Audio record & playback

Package type B (24 pins)

Pins		Function
(1)	Grounded during playback & E-E modes	
(2)	Grounded during record & E-E modes	
(3)	0v Ground	
(4)	0v	
(5)	0v	
(6)	0v5	
(7)	0v Ground	
(8)	3v2 →	Playback audio to pin 9 via playback level control
(9)	4v0 ←	From pin 8
(10)		
(11)	4v0 ←	450mV pk-pk Record mode audio from Tuner or external source
(12)	0v	
(13)	4v0 →	1v5 pk-pk audio output
(14)	0v	
(15)	0v	
(16)	1v5	
(17)	4v4 →	45v pk-pk record mode to ACE head
(18)	9v5	
(19)	←	12v Supply
(20)	←	LP mode control
(21)	0v	
(22)	←	Mute control
(23)	←	Mode control, E -E & Record mode = H, Playback = L
(24)	←	Record mode control

LA7285M

Audio record & playback **Package type B (24 pins)**

Pins		Function
(1) Playback mode grounded		
(2)	←	Playback from audio head
(3) 0v Ground		
(4)		
(5)		
(6) Playback level & equalization circuit		
(7)	→	Amplified playback to pin 8
(8)	←	From pin 7
(9)		
(10)	←	Audio from Tuner
(11) 0v Ground		
(12)	←	Audio from external source, i.e scart
(13)		
(14)	→	Audio output
(15)	←	Record level control
(16)		
(17)		
(18)	→	Audio to record head
(19)		
(20)	←	12v Supply
(21)	←	Speed control, LP = H, SP = L
(22)	←	Input source control. External = H, Tuner = intermediate Playback = L
(23)	←	Audio mute
(24)	←	Record mode = H, Playback = L

Audio record & playback amplifier

Package type B (30 pins)

Pins	Function
(1)	0v Ground, E-E & Playback modes
(2)	7v8
(3)	
(4)	
(5)	0v Ground, Record mode
(6)	0v5 ← Playback from ACE head
(7)	0v Ground, LP speed
(8)	0v Ground, EP speed
(9)	
(10)	0v Ground
(11)	2v7 → To playback level control
(12)	← From playback level control
(13)	← External Audio input
(14)	0v6
(15)	0v Ground
(16)	4v0 → Audio out
(17)	0v4
(18)	← Record audio
(19)	0v Ground, LP speed
(20)	0v Ground, EP speed
(21)	0v9
(22)	→ Record head
(23)	← LP speed switching control
(24)	← EP speed switching control
(25)	← Mute
(26)	← Playback = H, E-E = L control
(27)	← 9v Supply
(28)	8v4 → Oscillator bias
(29)	← Record = 9v0, playback = 0v control

LA7294

Audio processor

Package type A (30 pins)

Pins		Function
(1) 11v	→	Erase head bias oscillator drive
(2)		Record command = H
(3) 0v Ground		
(4)	→	E-E & Playback mode bias current switch off
(5) 8v		
(6) 9v		
(7) 0v Ground (LP mode)		
(8)	←	Switched 12v supply
(9)	→	Record or playback select switch
(10)	←	Playback mode command = H
(11) 0v Ground (LP mode)		
(12)	←	Mute command
(13) 0v Ground (LP mode)		
(14)	←	Tape speed LP command = H
(15)	←	Audio head playback
(16) 0v Ground		
(17)	←	LP Playback equalization circuit
(18)		
(19) 0v Ground		
(20) 4v	→	to audio head (record mode)
(21) 2v4	←	Audio head (playback signal)
(22) Record peaking circuit		
(23)	←	Playback signal from pin 21 via playback level & filter circuit
(24) 0v	←	Record level circuit
(25) 0v	←	Audio input signal from tuner or external
(26)		
(27) 0v4		
(28) 0v Ground (LP mode)		
(29)	←	Record & E-E mode automatic loudness control
(30) 4v4	→	Playback & record audio signal output

LA7297

Audio record & playback processor

Package type B (30 pins)

Pins		Function
(1)	⟶	Record 9v0 supply to bias circuit
(2)		
(3) 0v Ground		
(4) Grounded during playback mode		
(5) 8v0		
(6) 8v8		
(7)		
(8)	⟵	9v Supply
(9) Grounded during record mode		
(10)	⟵	H = Playback, L = Record
(11)		
(12)	⟵	H = audio mute
(13)		
(14)		
(15)	⟵	Playback audio from head
(16) 0v Ground		
(17)		
(18)	⟵	H = LP mode
(19) 0v Ground		
(20)	⟶	Record audio to head
(21) 2v4	⟶	Playback audio to pin 23
(22)		
(23)	⟵	From pin 21
(24)	⟵	Record mode audio from pin 30
(25)	⟵	Audio input from line & tuner
(26)		
(27)		
(28)		
(29)	⟵	Feedback from pin 30 for automatic loudness control
(30) 4v0	⟶	Playback & record modes audio output

LA7320

Record/playback head amplifier **Package type B (22 pins)**

Pins			Function
(1)		←	5v0 pk-pk head switching pulse
(2)		←	Playback & E-E mode 5v Supply
(3)	2v	→	FM output
(4)	0 v Ground		
(5)		←	Record Luminance signal
(6)		←	Record Chroma signal
(7)	0v Ground		
(8)	5v9 (PB), 0v (Rec')		
(9)			
(10)			
(11)		←	Record 12v Supply
(12)			
(13)		→	Record luminance & chroma signal to heads
(14)	0v Ground playback mode		
(15)	2v (PB)		
(16)		←	Head 1 (playback mode)
(17)	0V Ground		
(18)		→	Record current test terminal
(19)		←	Head 2 (playback mode)
(20)	2v (PB), 0v (Rec')		
(21)			
(22)			

LA7322

Luminance record & playback processing

Package type B (30 pins)

Pins		Function
(1)	←	FM Luminance input
(2)		Record mode Luminance pre-emphasis, RC & Set deviation circuit
(3)		Record mode Emphasis feedback RC circuit
(4)		Record mode Noise filter RC circuit. On = L, Off = H
(5)		Picture detail enhancement filter RC circuit, Record = L, Playback = H
(6)	←	500mV pk-pk video, Record mode luminance clamp circuit
(7)	→	Record mode = AGC to pin 6. Playback mode = demodulated signal to pin 9
(8)		0v Ground
(9)	←	500mV pk-pk video, Clamped
(10)	←	500mV pk-pk video, Playback from luminance delay line
(11)		Correlation detector test point
(12)	→	Playback luminance delay to pin 10
(13)		Noise de-emphasis RC circuit
(14)		Noise cancel filter RC circuit
(15)		Noise cancel filter RC circuit
(16)		Sync' separator RC filter
(17)	→	4v0 pk-pk Horizontal sync'
(18)	←	Picture control & Edit input, Edit on = H, Off = L
(19)	←	Playback chroma & E-E preset
(20)	←	Mute, Dummy vertical & horizontal pulse from servo
(21)	→	Playback & record Video output
(22)	←	5v Supply
(23)		
(24)	←	Record mode video
(25)		AGC circuit
(26)	←	400mV pk-pk, Playback FM luminance
(27)		AGC circuit
(28)	→	Drop out compensator detection
(29)	←	Head switching pulse, Tape speed LP =L, SP = H
(30)	←	1v5 pk-pk, Record FM luminance

LA7332

Chroma processor

Package type B (24 pins)

Pins		**Function**
(1) 0v5	←	Secam = H, NTSC = Medium, PAL = L
(2) 1v8		
(3) 3v0	←	400mV pk-pk Playback chroma from pin 5 via 4.43MHz filter
(4) 2v4		
(5) 2v5	→	400mv pk-pk Record & playback to pin 3 and pin 24 via 2 line delay
(6) 0v Ground		
(7) 3v0	←	400mV pk-pk, 627kHz playback chroma
(8) 2v8		
(9) 4v4	→	627kHz Record chroma to head circuit
(10) 2v8		
(11) 2v5	←	5.06MHz signal from sub convertor pin 13 via Bandpass filter
(12) 1v8		
(13) 4v5		180mV pk-pk, 5.06MHz Record & playback to main convertor pin 11
(14) 0v5	←	1v0 pk-pk Record & playback Horizontal sync' pulses
(15) 2v4		
(16) 4v7		
(17) 4v8		
(18) 2v1, 700mV pk-pk, 4.43MHz xtal oscillator		
(19) 1v7, 730mV pk-pk, 4.43MHz xtal oscillator		
(20)	←	5v Supply
(21) 1v7	←	520mV pk-pk Record mode chroma input
(22) 1v8	←	3v pk-pk, 25Hz flip flop pulse
(23) 1v9	→	520mV pk-pk, 4.43MHz playback chroma to luminance circuit
(24) 3v3	←	4.43MHz chroma from pin 5

LA7345

Luminance & Colour processor

Package type C (44 pins)

Pins		Function
(1)	←	LP mode = H
(2)		
(3)		
(4)	←	Luminance from drop out compensator delay line
(5) n/c		
(6) 0v Ground		
(7)	→	Luminance to delay line
(8)	←	Picture sharpness control
(9)		
(10)	→	629kHz Record to head preamplifier
	←	629kHz Playback chroma from head preamplifier
(11)	←	Playback mode = H
(12)	←	Still & slow speed compensation
(13) 4.43MHz xtal oscillator		
(14)	→	4.43 MHz clock pulse to servo circuits
(15)		
(16)	→	8.86MHz clock pulse to delay line
(17)		
(18)		
(19)		
(20)	→	Playback chroma to pin 22 via delay line & 5.7MHz trap circuits
(21)		
(22)	←	From pin 21
(23)	←	5v Supply
(24) n/c		
(25)	→	Internally generated regulated 4v5 for Luminance AGC control
(26) n/c		
(27)	→	Composite sync' pulse to servo circuits
(28)	←	Vertical lock pulses
(29) n/c		
(30)	→	Playback video output
(31)	←	5v Supply
(32)	→	Burst gating pulse
(33)	←	Video from Tuner or AV source
(34)		

(Continued overleaf)

LA7345

(35)	←	Playback luminance FM from head circuit
(36)	→	Record mode Luminance FM
(37)	←	25HzHead switching pulse
(38)	←	FM modulator deviation & carrier controls
(39) n/c		
(40)	←	Playback level control
(41)	→	Playback & Record Luminance to pin 42 via 3.5MHz low pass filter
(42)	←	From pin 41
(43)		
(44)		

Note :
Pin 32 may not be connected

LA7376

Video head playback & record amplifier

Package type B (16 pins)

Pins		Function
(1)	←—	5v Supply
(2) 2v8	—→	FM envelope detection
(3) 0v3	←—	25Hz Head switching pulse
(4) 2v0	—→	250mV pk-pk Playback chroma
(5) 0v Ground		
(6) 2v5	—→	300mV pk-pk Playback FM output
(7) 1v0		
(8) 2v0	←—	Record mode Luminance & chroma signals
(9) 2v0	—→	Record mode mixed luminance & chroma to pin 10
(10) 1v6	←—	From pin 9
(11)	—→	Record mode, 4v5 pk-pk Luminance & chroma to video heads
(12) 0v7	←—	Playback FM from head 1
(13)	←—	Mode select, Record = 4v, Playback = 0v
(14) 0v Ground		
(15)	←—	Mode select, Record = 4v, Playback = 0v
(16) 0v7	←—	Playback FM from video head 2

Note :
Pin 9 sometimes not used, Luminance & chroma record signal goes straight to pin 10

LA7377

Head record & playback amplifier

Package type B (22 pins)

Pins		Function
(1)	←	5v Supply
(2)	←	5v Supply
(3) n/c		
(4)	→	FM envelope detection
(5)	←	Head flip flop switching pulse
(6)	→	Playback FM luminance to pin 8 via equalization circuit & 629Khz chroma to low pass filter circuit
(7) 0v Ground		
(8)	←	Luminance from pin 6
(9)		
(10)	→	Playback Luminance FM
(11)	←	Record mode Luminance & chroma
(12)	→	500mV pk-pk Record mode Luminance & chroma to pin 13
(13)	←	Record mode Luminance & chroma from pin 12
(14) n/c		
(15)	→	Record mode FM to heads, grounded during playback
(16)	←	Playback FM from head 2
(17)		
(18) 0v Ground		
(19)		
(20) n/c		
(21)	←	Playback FM from head 1
(22) n/c		

Note :
Pin 3, 14, 20 & 22 may be connected to ground

LA7390

Luminance & chroma processor

Package type B (36 pins)

Pins		Function
(1)		
(2)	⟶	0v5 pk-pk Composite video to pin 3
(3)	⟵	Composite video from pin 2
(4)		
(5)		
(6) 4v5		
(7) 2v5		
(8) 3v6		
(9) 2v0	⟵	From drop out compensation & Luminance delay circuit
(10) 0v Ground		
(11) 2v0	⟶	To pin 9
(12)	⟵	Playback picture control
(13) 1v7		
(14) 3v0	⟶	300mV pk-pk, 627kHz record chroma
	⟵	150mV pk-pk playback chroma
(15) 2v4		
(16) 2v2		
(17) 4v0, 4.43MHz xtal		
(18) 4.43MHz, 0v5 pk-pk xtal		
(19) 3v3		
(20) 3v3		
(21) 2v4	⟶	Playback Chroma to pin 23 via comb filter
(22) 2v0		
(23) 2v0	⟵	Chroma from comb filter
(24)	⟵	5v Supply
(25)	⟶	Regulated 4v2 supply
(26) 0v6	⟵	4v0 pk-pk Composite sync' pulse
(27) 0v1	⟵	External character insertion
(28) 2v2	⟶	Composite video, 2v0 pk-pk Playback, 2v2 pk-pk Record
(29)	⟵	5v Supply
(30)		
(31) 3v4	⟵	1v2 pk-pk Composite video from Tuner or external
(32) 1v5		
(33) 3v0	⟵	0v4 pk-pk Playback FM from video head circuit
(34)	⟶	1v0 pk-pk FM to record video heads
(35) 0v8	⟵	Drum flip flop
(36) 2v5	⟵	FM modulator frequency & deviation controls

LA7391

NTSC, PAL, SECAM luminance & chroma processor

Package type B (42 pins)

Pins		**Function**
(1) 0v1	←	Drop out compensation pulse
(2) 2v1	←	Luminance playback level control
(3) 1v7	→	800mV pk-pk record luminance
(4) 2v4	←	From pin 3
(5) 4v8	→	1v0 pk-pk record luminance to pin 42
(6) 5v0		
(7) 5v0		
(8) 2v5		
(9) 3v7		
(10) 2v1	←	400mV pk-pk playback luminance from delay line
(11) 0v Ground		
(12) 2v0	→	500mV pk-pk playback luminance to delay line
(13) 2v4	←	Picture control
(14) 1v4		
(15) 2v8	←	100mV pk-pk playback chroma from head circuit
	→	400mV pk-pk record chroma to head circuit
(16) 2v4		
(17) 2v1	←	Trick speed indication from servo circuit
(18) 3v8, 4.43MHz xtal		
(19) 2v6, 4.43MHz xtal		
(20) 1v6		
(21) 5v2	→	8.86MHz clock pulse to delay line
(22) 0vGround		
(23) 3v2		
(24) 3v4, trick speed modes 0v		
(25) 3v0	→	2v0 pk-pk, 4.43MHz playback chroma to pin 27 via bandpass filter
(26) 2v0		
(27) 2v5	←	Playback 4.43MHz chroma from pin 25
(28)	←	5v Supply
(29) 0v	←	Playback 4.43MHz chroma from Mesecam circuit
(30) 4v1		
(31) 2v7	→	Playback 4.43MHz chroma to Mesecam circuit
(32) 0v4	→	Vertical sync' pulse
(33) 0v2	←	Vertical sync' pulse
(34) 2v1	→	2v0 pk-pk video output
(35)	←	5v Supply

(Continued opposite)

(36) 1v5	——→	Burst gating pulse, PAL / SECAM sensing
(37) 3v3	←——	1v0 pk-pk record mode video
(38) 1v4		
(39) 3v2	←——	250mV pk-pk playback FM
(40) 3v8	——→	800mV pk-pk record luminance FM to head circuit
(41) 0v7		
(42) 0v	←——	50mV pk-pk record luminance luminance from pin 5

LA7505

Sound & vision I.F detector

Package type B (16 pins)

Pins		Function
(1)	←	I.F from SAW Filter
(2)		
(3)	←	RF, AGC adjustment
(4)	→	Tuner AGC control
(5) AFT LC circuit		
(6) n/c		
(7)	→	AFT to Tuner
(8) Vision demodulator LC circuit		
(9) Vision demodulator LC circuit		
(10) AFT, LC circuit		
(11)		
(12)	→	Composite video output & sound to I.F circuit
(13) 0v Ground		
(14)		
(15)		
(16)	←	I.F from SAW Filter

LA7522

Vision & sound I.F demodulator

Package type B (30 pins)

Pins		Function
(1)	Sound FM detector circuit	
(2)	——→	Audio output
(3)	←——	Audio amplifier input
(4)		
(5)	——→	Audio amplifier output
(6)	0v Ground	
(7)		
(8)		
(9)	←——	I.F from SAW Filter
(10)	←——	I.F from SAW Filter
(11)		
(12)	←——	Tuner AGC manual adjustment control
(13)	——→	AGC control to tuner
(14)		
(15)	n/c	
(16)	n/c	
(17)	——→	AFC control to tuner
(18)	AFC, LC circuit	
(19)	Video detector LC circuit	
(20)	Video detector LC circuit	
(21)	AFC, LC circuit	
(22)	——→	Video output
(23)	←——	12v Supply
(24)	←——	12v Supply
(25)	——→	Sound I.F output to pin 27 via filter
(26)		
(27)	←——	Sound I.F input from pin 25
(28)		
(29)	——→	Audio mute to amplifier on pins 1,2 & 3
(30)	Sound FM detector circuit	

Alternative :
Pins 3,4 & 5 contain a separate amplifier. If not used, pin 3 grounded, pins 4 & 5 n/c
Pin 25 n/c

LA7530N

I.F amplifier detector

Package type B
(20 pins)

Pins		Function
(1) 5v6		
(2) 4v6		
(3) 4v9	→	Audio output
(4) 3v6		
(5) 3v6		
(6) 4v3	←	I.F from SAW Filter
(7) 4v3	←	I.F from SAW Filter
(8) 0v Ground		
(9) 0v9	←	AGC control
(10) 3v1	→	AGC to tuner
(11) 0v3		
(12) 3v8 AFT, LC circuit		
(13) 8v2 Vision detector LC circuit		
(14) 8v2 Vision detector LC circuit		
(15) 3v7, AFT LC circuit		
(16) 4v7	→	Video output
(17)	←	12v Supply
(18) 0v Ground		
(19)	←	Sound I.F from pin 16
(20) 2v9	←	Audio mute control

LA7555

Vision & sound I.F detector

Package type B (24 pins)

Pins		Function
(1) 6v5	⟶	Audio output
(2) 3v0		
(3)		
(4)		
(5)		
(6)	⟵	12v Supply
(7) 0v Ground		
(8) 4v5	⟵	I.F from SAW Filter
(9) 4v5	⟵	I.F from SAW Filter
(10) 4v0	⟵	AGC control
(11) 11v3	⟶	AGC control to tuner
(12) 12v0		
(13) 12v0		
(14)		
(15)		
(16)		
(17)	⟵	12v Supply
(18) 10v5		
(19) 10v5		
(20) 5v5		
(21) 5v6	⟶	Video output
(22) 0v		
(23)		
(24) 2v0		

LA7913

Tuning & band selection interface

Package type B (16 pins)

Pins		Function
(1)	←	13v Supply
(2)	←	Tuning band control logic
(3)	←	Tuning band control logic
(4)	←	Tuning band control logic
(5) 0v Ground		
(6)	←	33v Supply
(7)	←	Pulse width modulated tuning input
(8)	→	Tuning voltage output
(9)	←	Tuning voltage feedback
(10)		
(11) n/c		
(12)	→	UHF band select
(13)	→	VHF H select
(14)	→	VHF L select
(15) n/c		
(16)	←	12v Supply

Pin 4 may be n/c

LA7925

Audio playback & record

Package type A (30 pins)

Pins		Function
(1)	←	Record control = L
(2)	←	Record start control, Record = L, E-E = H
(3) 0v Ground		
(4) Grounded during Playback mode		
(5) 8v0		
(6) 9v1		
(7) 0v		
(8)	←	12v Supply
(9) 0v Ground		
(10)	←	Mode switching, Playback = H, Record = L
(11) 0v		
(12)	←	Audio mute = H
(13) 2v2		
(14)	←	NTSC mode, SP = L, EP = H
(15) 2v2	←	Playback from audio head
(16) 0v Ground		
(17) 2v2		
(18) 0v		
(19) 0v Ground		
(20) 4v0	→	Record audio to head
(21)	→	100mV pk-pk amplified playback from pin 15
(22) 1v3		
(23)	←	From pin 21
(24)	←	Record mode audio from pin 30
(25)	←	E-E / AUX' record mode audio
(26)		
(27) 0v		
(28)		
(29)		
(30) 3v0	→	2v5 pk-pk audio output

LA7934

Tuning band selection

Package type B (20 pins)

Pins		Function
(1) 0v6		
(2) 0v		
(3)	→	No signal video mute output
(4)	←	Band select data
(5)	←	Band select data
(6) 0v Ground		
(7)	←	Video mute = Hi
(8) 6v4		
(9)	→	Power on reset = Lo
(10)	→	Regulated 5v6 Supply
(11)	←	9v Supply
(12)	←	Sync' input = Lo
(13)		
(14)		
(15)	→	Mute
(16)	←	12v Supply
(17)	→	Selected tuning band
(18)	→	Selected tuning band
(19)	→	Selected tuning band
(20)	→	Selected tuning band

LA7935

Tuning band selection

Package type B (22 pins)

Pins		**Function**
(1) 0v2	←	Video sync' pulse input
(2) 5v5	→	Inverted sync' pulses to pin 3
(3)	←	From pin 2
(4) 5v0	→	Station detected (sync' pulse) signal
(5)	←	Tuning band selection data
(6)	←	Tuning band selection data
(7) 0v Ground		
(8) 6v2		
(9) 5v0		
(10)	←	Power on reset = L
(11) 0v		
(12)	→	Regulated 5v6 supply
(13)	←	12v Supply
(14)	←	40v Supply
(15)	→	32v Supply
(16) 6v7	←	AFC control
(17)	→	AFC output
(18)	←	12v Supply
(19)	→	VHF L band
(20)	→	VHF H band
(21) n/c		
(22)	→	UHF band

LA7938

Tuning band interface

Package type B (22 pins)

Pins		Function
(1)	←	Composite sync' input
(2)	→	Inverted sync to pin 3
(3)	←	From pin 2
(4)	→	Detected sync' pulses to system control
(5)	←	VHF & UHF bands select control
(6)	←	VHF & UHF bands select control
(7) 0v Ground		
(8)	←	AFC control from pin 17
(9) Regulated 5v0		
(10)	←	Power on reset = L
(11)		
(12) Regulated 5v6		
(13)	←	9v Supply
(14)	←	36v unregulated tuning supply
(15)	→	32v Regulated tuning supply
(16)	←	AFC from I.F circuit
(17)	→	To pin 8
(18)	←	9v Supply
(19)	→	VHF, UHF selected Band to tuner
(20)	→	VHF, UHF selected Band to tuner
(21)	→	VHF, UHF selected Band to tuner
(22)	→	VHF, UHF selected Band to tuner

LB1616

Capstan motor drive

Package type B (20 pins)

Pins		Function
(1)	→	Stator coil 1 drive
(2)	→	Stator coil 2 drive
(3)	→	Stator coil 3 drive
(4)	←	12v Supply
(5)		
(6)		
(7)		
(8) 0v Ground		
(9)	←	5v Supply
(10) 0v motor ground		
(11)	←	Capstan motor control from servo
(12)	←	Forward = H, Reverse = L
(13)	←	+ Hall effect device 3
(14)	←	– Hall effect device 3
(15)	←	+ Hall effect device 2
(16)	←	– Hall effect device 2
(17)	←	+ Hall effect device 1
(18)	←	– Hall effect device 1
(19)	←	5v Supply
(20)		

LB1620

Drum motor drive | **Package type B (20 pins)**

Pins		Function
(1) 5v8	⟶	6v1 pk-pk drive to stator coil 1
(2) 5v8	⟶	6v1 pk-pk drive to stator coil 2
(3) 5v8	⟶	6v1 pk-pk drive to stator coil 3
(4)	⟵	Switched 12v Supply
(5) 0v9		
(6) 0v9		
(7) 0v9		
(8) 0v Ground		
(9)	⟵	Switched 9v supply
(10) 0v		
(11) 4v4	⟵	Control voltage from servo
(12) 0v Ground		
(13) 3v4	⟵	Hall effect device 1
(14) 3v4	⟵	Hall effect device 1
(15) 4v6	⟵	Hall effect device 2
(16) 4v6	⟵	Hall effect device 2
(17) 5v8	⟵	Hall effect device 3
(18) 5v8	⟵	Hall effect device 3
(19)	⟵	Switched 9v supply
(20) 0v		

LB1641

Loading motor drive

Package type A (10 pins)

Pins		Function
(1)	0v Ground	
(2)	⟶	–ve drive to motor
(3)	Capacitor coupled to pin 9	
(4)	⟵	H / L speed Function (torque) commands from system control
(5)	⟵	Motor reverse command
(6)	⟵	Motor forward command
(7)	⟵	13v Supply
(8)	⟵	12v Supply
(9)	Capacitor coupled to pin 3	
(10)	⟶	+ve drive to motor

LB1649

2 × Independent Motor drivers

Package type B (12 pins)

Pins		Function
(1)	←	15v Supply Motor driver 2.
(2) 8v8	←	Motor driver 2. Forward, reverse & brake mode control
(3) 0v5	→	Drive to motor 2
(4) 0v5	→	Drive to motor 1
(5) 8v8	←	Motor driver 1. Forward, reverse & brake mode control
(6) 0v3		
(7)	←	15v Supply Motor driver 1.
(8) 8v8	←	Motor driver 1. Forward, reverse & brake mode control
(9) 0v5	→	Drive to motor 1
(10) 0v5	→	Drive to motor 2
(11) 8v8	←	Motor driver 2. Forward, reverse & brake mode control
(12) 0v3		

Typically used for carriage & loading control functions

LB1684

Drum motor drive

Package type B (20 pins)

Pins			Function
(1)	7v0	→	Stator coil 1 drive
(2)	7v0	→	Stator coil 2 drive
(3)	7v0	→	Stator coil 3 drive
(4)		←	16v Supply
(5)	0v8		
(6)	0v8		
(7)	0v8		
(8)	0v Ground		
(9)	0v7		
(10)		→	Output current sensing & feedback to pin 20
(11)	2v6	←	Drive from servo
(12)	0v		
(13)	2v6	←	+ve Hall effect device 3
(14)	2v6	←	–ve Hall effect device 3
(15)	2v6	←	+ve Hall effect device 2
(16)	2v6	←	–ve Hall effect device 2
(17)	2v6	←	+ve Hall effect device 1
(18)	2v6	←	–ve Hall effect device 1
(19)		←	5v Supply
(20)		←	From pin 10

LB1807

Capstan Motor drive

Package type B (28 pins)

Pins		Function
(1)	⟶	Drive to stator coil 1
(2)	⟶	Drive to stator coil 2
(3)	⟶	Drive to stator coil 3
(4)		
(5)	⟵	+ve Hall effect device 3
(6)	⟵	–ve Hall effect device 3
(7)	⟵	+ve Hall effect device 2
(8)	⟵	–ve Hall effect device 2
(9)	⟵	+ve Hall effect device 1
(10)	⟵	–ve Hall effect device 1
(11) 0v Ground		
(12)		
(13) Motor 0v Ground		
(14)		
(15) Motor 0v Ground		
(16)	⟵	Speed & phase control
(17)	⟵	Forward = H, Reverse = L
(18)		
(19) 0v Ground		
(20)	⟶	1 v pk-pk FG pulses to servo circuit
(21)	⟵	FG pulses
(22)	⟵	FG pulses
(23)		
(24)	⟵	5v Supply
(25)	⟵	13v Motor power supply
(26)		
(27)		
(28)		

LC66508B

System control microprocessor

Package type B (64 pins)

Pins		**Function**
(1)	←	Serial Data
(2)	→	Serial Data
(3)	→	Serial Clock
(4)	←	30 Hz pulse during play back & record modes
(5)	←	30 Hz pulse synchronized to E-E video or to the head switching during playback
(6)		
(7)		
(8)	←	5v Supply monitoring
(9)	←	Mode switch
(10)	←	Mode switch
(11)	←	Mode switch
(12)	←	Anti - record switch (Tape tab sensor)
(13)		
(14)	←	Search mode control pulse from servo circuits
(15)	→	Servo control amplifier gain switch
(16)	→	Capstan motor brake
(17)	←	Data from servo auto tracking circuits
(18)	→	Tracking data to servo circuits
(19)	→	Serial clock to Servo circuits
(20)		
(21)		
(22)		
(23)		
(24)		
(25)		
(26)		
(27)		
(28)		
(29)	←	30 Hz playback head switching pulse
(30) 0v Ground		
(31) 0v Ground		
(32) 4MHz clock		
(33) 4MHz clock		
(34)	←	Power on reset
(35)	←	Still mode noise shift
(36)	←	AV input = L, Aerial input = H

(Continued overleaf)

LC66508B

(37)	←	Capstan FG pulse
(38)	→	Forward direction loading Motor drive
(39)	→	Reverse direction loading motor drive
(40)	→	Slow speed loading motor drive
(41)	→	Medium speed loading motor drive
(42)	→	H = Record mode command to video circuits
(43)	→	H = Playback mode command to video circuits
(44)	→	H = Still,SLOW & X2 speed modes to video circuits
(45)	→	Cue, reverse, still, slow, frame advance, audio dub & pause modes
(46)	→	H =Aerial, L = AV
(47)		
(48)		
(49)	←	Supply voltage reference
(50)	←	Take up reel pulses
(51)	←	Dew sensor
(52)	←	Tape end sensor
(53)	←	Tape end sensor
(54)	←	Front loading, L = tape out, Medium = Tape loading, H = Tape loaded
(55)	←	Tape speed, H = LP, L = SP
(56)	→	Audio mute
(57)	→	Audio record mute
(58)		
(59)		
(60)		
(61)	→	Power on to power supply circuit
(62)	→	Audio stereo / left /right select
(63)	→	Audio stereo / left /right select
(64)	←	5v Supply

LA7417

Digital Servo

Package type C (48 pins)

Pins		Function
(1)	→	Head switching control for 3 head & 4 head machines
(2)	→	Head switching control for 3 head & 4 head machines
(3)	→	20 millisec' Head switching pulse
(4)	→	20 millisec' FM audio switching pulse
(5)	←	Vertical lock preset
(6)		
(7)	←	3v8 pk-pk Control pulse from audio /control head
(8)	→	Recording control, H, M, L
(9)	→	Recording control, H, Open, L
(10)	→	Slow mode Drum compensation pulse
(11)	←	Drum PG delay preset
(12)	←	3v9 pk-pk, 40 msec' Drum PG
(13)	←	4v0 pk-pk, 1.3 msec', Drum FG
(14)	→	5v0 pk-pk, 8.59 kHz, Drum speed control error digital voltage
(15)	→	Horizontal sync' (frequency to digital voltage converted), taken from the composite video waveform
(16)	→	5v0 pk-pk, 8.59 kHz, Drum phase control error digital voltage
(17)	→	Slow mode Capstan compensation pulse
(18)	→	Slow mode Capstan brake & start pulses
(19)	←	5v Supply
(20)	→	5v0 pk-pk, 17 kHz, Capstan speed error digital voltage
(21)	→	5v0 pk-pk, 8.59 kHz, Capstan phase error digital voltage
(22)	→	Capstan forward mode = L, Reverse mode = H
(23)	←	4v0 pk-pk, 0.6 msec', Capstan FG
(24)	←	Capstan FG (used to detect capstan stop during slow mode)
(25)	→	Recording modes, EP =H, LP = Medium, SP = L
(26)	→	Recording mode, SP = L
(27)	→	Recording mode, SP = H
(28)	←	Capstan forward or reverse control

(29) Slow tracking adjustment circuit
(30) Tracking adjustment circuit
(31) Slow mode Drum timing compensation circuit
(32) Slow mode Drum timing compensation circuit

(Contined overleaf)

LA7417

(33) Slow mode Drum timing compensation circuit
(34) ——→ Tracking delay signal
(35) ——→ Slow mode = H, Still mode = L
(36) ——→ 2 × speed & start of slow / still modes = H
(37) ——→ 5v0 pk-pk, 40 msec',Clock control timing signal
(38) ——→ Control signal data
(39) ——→ Serial clock
(40) ——→ Data
(41) ←—— NTSC mode = H, PAL mode = L
(42) ——→ 20 msec', Internal reference signal
(43) ←—— 5v Supply
(44) ←—— 1v0 pk-pk, 4.43MHz Colour subcarrier signal (for internal reference clock generator)
(45) 0v Ground
(46) ←—— Composite video
(47) ←—— 4 Head special playback envelope detection signal
(48) ——→ Vertical lock pulse

The following pins may not be connected (n/c)
1.2.4.6.10.15.17.18.24.27.29.31.32.33.34.35.36.
Pin 1. may be connected to a switched 5v supply
pin 47 may be connected to 0v ground

LC74760

On screen display control

Package type B (30 pins)

Pins		**Function**
(1) 0v Ground		
(2) n/c		
(3) n/c		
(4)	◄────	Horizontal sync' pulse
(5) 17.7 MHz xtal oscillator		
(6) 17.7 MHz xtal oscillator		
(7)	◄────	Vertical sync' pulse
(8)	◄────	Enable signal from system control
(9)	◄───►	Data
(10)	◄────	System clock pulse from system control
(11) n/c		
(12) n/c		
(13) n/c		
(14)	────►	Picture tone control
(15)		
(16)	────►	Composite video
(17)	◄────	5v Supply
(18)	◄────	Video from picture tone circuit
(19) n/c		
(20)	◄────	Video sync' pulses
(21) n/c		
(22) 0v Ground		
(23) 2v0		
(24) 2v0		
(25) 2v5		
(26) 2v5		
(27) 2v1		
(28) 2v2		
(29) n/c		
(30)	◄────	5v Supply

Note :
Pin (4) may be n/c
Pin (7) may be n/c

LC7530

Electronic Volume and display control

Package type B (20 pins)

Pins		Function
(1)	——→	LED Display
(2)	——→	LED Display
(3)	——→	LED Display
(4)	——→	LED Display
(5)	——→	LED Display
(6)	——→	LED Display
(7)	——→	LED Display
(8)	——→	LED Display
(9) 4v4	←——	Right channel
(10) 4v4	——→	Right channel output
(11) 4v4		
(12) 0v Ground		
(13) 4v4	——→	Left channel output
(14) 4v4	←——	Left channel
(15)	←——	Enable (normally 12 Supply)
(16)		
(17)	←——	Volume down control = L
(18)	←——	Volume up control = L
(19) 0v		
(20)		

LC8992

Luminance delay line

Package type B
(8 pins)

Pins			Function
(1)		←	9v Supply
(2)	0v Ground		
(3)			
(4)	3v4	→	400mV pk-pk delayed luminance
(5)	4v6		
(6)	3v2	←	400mV pk-pk luminance
(7)	2v0	←	8.85MHz, 4v0 pk-pk clock pulse
(8)	2v8		

Note :
Above example as used in drop out compensation

LC89925

Luminance Delay line

Package type B
(8 pins)

Pins		Function
(1) n/c or Ground		
(2) 0v Ground		
(3)	←	5v Supply
(4)	→	Delayed video
(5) n/c		
(6)	←	Video
(7)	←	8.86MHz clock pulse
(8)		

Note:
Above example as used in drop out compensation

LM324

4 × Operational amplifiers

Package type B (14 pins)

Pins		Function
(1)	⟶	Amplifier 1 output
(2)	⟵	Amplifier 1, inverted input
(3)	⟵	Amplifier 1, non inverted input
(4)	⟵	32v Supply
(5)	⟵	Amplifier 2, non inverted input
(6)	⟵	Amplifier 2, inverted input
(7)	⟶	Amplifier 2 output
(8)	⟶	Amplifier 3 output
(9)	⟵	Amplifier 3 inverted input
(10)	⟵	Amplifier 3 non inverted input
(11) 0v Ground		
(12)	⟵	Amplifier 4, non inverted input
(13)	⟵	Amplifier 4, inverted input
(14)	⟶	Amplifier 4 output

Amplifier designated numbers, i.e 1,2, are only given to distinguish between the individual devices

LM339N

4 × Operational amplifiers **Package type B (14 pins)**

Pins		Function
(1)	⟶	Amplifier 3
(2)	⟶	Amplifier 4
(3)	⟵	12v Supply
(4)	⟵	Amplifier 4 inverted input
(5)	⟵	Amplifier 4 non inverted input
(6)	⟵	Amplifier 3 inverted input
(7)	⟵	Amplifier 3 non inverted input
(8)	⟵	Amplifier 1 inverted input
(9)	⟵	Amplifier 1 non inverted input
(10)	⟵	Amplifier 2 inverted input
(11)	⟵	Amplifier 2 non inverted input
(12) 0v Ground		
(13)	⟶	Amplifier 2 output
(14)	⟶	Amplifier 1 output

Amplifier designated numbers, i.e 1,2, etc are only given to distinguish between the individual devices

LM358

Dual operational amplifier

Package type B (8 pins)

Pins		**Function**
(1)	⟶	Amplifier 1 output
(2)	⟵	Amplifier 1 Inverted input
(3)	⟵	Amplifier 1 non inverted input
(4)	0v Ground	
(5)	⟵	Amplifier 2 non inverted input
(6)	⟵	Amplifier 2 inverted input
(7)	⟶	Amplifier 2 output
(8)	⟵	12v Supply

Amplifier designated numbers, i.e. 1,2 are only given to distinguish between the individual devices

LM393P

Dual operational amplifiers **Package type B (8 pins)**

Pins		**Function**
(1)	⟶	Amplifier 1 output
(2)	⟵	Amplifier 1 non inverted input
(3)	⟵	Amplifier 1 inverted input
(4)	0v Ground	
(5)	⟵	Amplifier 2 non inverted input
(6)	⟵	Amplifier 2 inverted input
(7)	⟶	Amplifier 2 output
(8)	⟵	5v Supply

Amplifier designated numbers, i.e 1,2, etc are only given to distinguish between the individual devices

LVA516S

Sync' pulse separator with automatic frequency control

Package type A (9 pins)

Pins			Function
(1)	2v5	←	Luminance video input
(2)	2v3		
(3)	2v3		
(4)	0v7		
(5)	0v Ground		
(6)	3v2		
(7)	4v6	→	Vertical sync' pulse output
(8)	4v6	→	Horizontal sync' pulse output
(9)		←	5v Supply

LVA521

2 × input to 1 output, amplified buffered selector switch

Package type A (9 pins)

Pins		Function
(1)		
(2)	←	9v Supply
(3)		
(4)	→	Amplified output
(5) 0v Ground		
(6)	←	Audio input 1
(7)	←	Select control, Audio 1 output = L, Audio 2 output = H
(8)	←	Audio input 2
(9) 0v Ground		

LVA873B

RF convertor **Package type B (12 pins)**

Pins		Function
(1)	⟶	Modulated video output
(2)	⟶	Regulated supply to oscillator
(3)	⟶	FM sound output
(4) n/c		
(5) n/c		
(6) 0v Ground		
(7) n/c		
(8)	⟵	From pin 2
(9)	⟵	Audio input
(10)	⟵	5v Supply
(11) 0v Ground		
(12)	⟶	Video input

M35011

On screen display generator

Package type B (20 pins)

Pins	Function
(1) 2v7, Character position LC circuit	
(2) 2v6, Character position LC circuit	
(3)	← Enable
(4)	← Serial clock pulse
(5)	← On screen display data
(6)	← Reset = L
(7)	← 6v Supply
(8)	→ Video
(9)	← Character level control
(10) 5v0	← Video
(11) 0v Ground	
(12) 0v	
(13) 1v0	
(14) 5v0	
(15) 5v0	
(16) 5v0, 17.7 MHz xtal oscillator	
(17) 5v0, 17.7 MHz xtal oscillator	
(18)	← Horizontal sync' pulse
(19)	← Vertical sync' pulse
(20)	← 6v Supply

M50115AP

Vision & sound I.F detector

Package type B (30 pins)

Pins		Function
(1) 3v9	←	RF, AGC control
(2) 4v5		
(3) 4v4		
(4)	←	12v Supply
(5) 0v5		
(6) 0v Ground		
(7) 4v0		
(8) 4v1	←	I.F from SAW Filter
(9) 4v1	←	I.F from SAW filter
(10) 4v1		
(11) 0v Ground		
(12)	←	Audio attenuator
(13)	←	12v Supply
(14) 7v0 FM Detector 6MHz xtal circuit		
(15) 5v0 FM Detector 6MHz xtal circuit		
(16) 5v7		
(17) 5v0	→	Audio output
(18) 5v0	←	Sound I.F from 6MHz xtal filter
(19) 5v0		
(20) 2v8	→	Vision I.F to 6MHz xtal filter
(21) 6v0		
(22) 3v0 Sound trap 33.5 MHz CL circuit		
(23) 3v0 Sound trap 33.5 MHz CL circuit		
(24) 7v0	→	Automatic frequency control to tuner
(25) 3v7 AFC, CL circuit		
(26) 6v1 Video detector circuit		
(27) 6v1 Video detector circuit		
(28) 3v8		
(29) 7v6	→	Video output
(30) 3v2	→	Automatic gain control to tuner

M50127AP

Serial to 8 bit parallel convertor

Package type B (18 pins)

Pins		**Function**
(1)	0v Ground	
(2)	455 kHz xtal oscillator	
(3)	455 kHz xtal oscillator (9v0 pk-pk)	
(4)	——→	Data output
(5)	——→	Data output
(6)	——→	Data output
(7)	——→	Data output
(8)	——→	Data output
(9)	——→	Data output
(10)	——→	Data output
(11)	——→	1 Bit H or L output
(12)	n/c	
(13)	0v Ground	
(14)	——→	Data output
(15)	n/c	
(16)	←——	Serial data input
(17)		
(18)	←——	10v Supply

M50430

Channel selection & tuning processor

Package type B (30 pins)

Pins		Function
(1)	←	5v Supply
(2)	→	Clock pulses to memory i.c
(3)	↔	Read, write, erase data
(4)	↔	Read, write, erase data
(5)	↔	Read, write, erase data
(6)	←	Power on reset = L
(7)	→	Channel data
(8)	→	Channel data
(9)	4MHz oscillator circuit	
(10)	4MHz oscillator circuit	
(11)	↔	Memory data
(12)	←	Channel clock pulses
(13)	←	Channel position data
(14)	←	Key scan line return
(15)	←	Key scan line return
(16)		
(17)		
(18)	→	Pulse width modulated tuning
(19)	→	Memory i.c "select" command
(20)	→	AFC defeat = L
(21)	→	Inter-station audio mute = H
(22)		
(23)	→	Key scan pulses
(24)	→	Key scan pulses
(25)	→	Key scan pulses
(26)		
(27)		
(28)	→	Band switching control, VHF L = L, VHF H = H, UHF = L
(29)	→	Band switching control, VHF L = L, VHF H = L, UHF = H
(30)	→	Band switching control, VHF L = H, VHF L = H, UHF = L

M50455-084SP

On screen character generator **Package type B (32 pins)**

Pins		Function
(1) 0v Ground		
(2) 5v0	←	Clock pulse
(3) 4v9	←	Power down memory = L
(4) 2v4, Oscillator		
(5) 2v4	←	Charactors screen position
(6)	←	PAL = L, NTSC = H
(7) 3v3	←	Enable
(8) 0v4	←	Data
(9) 0v		
(10) 5v0		
(11) 0v Ground		
(12) 1v0	→	Composite video
(13) 1v0	←	Composite video
(14) 0v5		
(15) 2v2		
(16) 0v5		
(17) 1v4		
(18)	←	5v Supply
(19) 0v3		
(20) 0v3		
(21)		
(22)	→	Blanking pulse
(23)		
(24)	→	Blue - Luminance
(25)	→	Green - Luminance
(26)	→	Red - Luminance
(27)	→	Composite sync' pulse
(28)		
(29) 2v2	←	Oscillator control
(30) 0v5	←	Horizontal sync' pulse
(31) 4v7	←	Vertical sync' pulse
(32)	←	5v Supply

Pins 21 to 28 may not be used

M50730-610

Control processor

Package type B
(52 pins)

Pins		Function
(1)	→	Power on / off control, On = L
(2)	→	Audio mute control
(3)	→	Playback = L, E-E = H
(4)	→	Record 12v mode on = L
(5)	→	Vertical pulse for pause & search modes = H
(6)	→	Playback search mode = H
(7)	→	Servo motor start = H
(8)		
(9)	←	Drum motor flip flop pulse
(10)	→	Record mode on = L
(11)		
(12)	↔	Data
(13)	→	Clock pulse
(14)	←	Take up reel FG pulse
(15)	←	Input select, Tuner = L, Other = H
(16)		
(17)	←	Power on / off command
(18)	←	Capstan FG pulse
(19)	←	Remote camera pause mode on =L
(20)	←	Timer on = L
(21)	←	Remote control data
(22) 0v Ground		
(23)	←	Reset = L
(24) 4MHz clock xtal		
(25) 4MHz clock xtal		
(26)		
(27) 0v Ground		
(28)	←	Tape start sensor on = L
(29)	←	Tape end sensor on = L
(30)	←	Anti record switch on = L
(31)	←	Tape in switch on = L
(32)	←	Tape loaded sensor switch on = L
(33)	←	Tape ejected sensor on = L
(34)	←	Deck position mode sensor
(35)	←	Deck position mode sensor
(36)	→	Mode motor control
(37)	→	Mode motor control

(Continued overleaf)

M50730-610

(38) ⟶ Mode motor control
(39) ⟶ Mode motor control
(40) 0v Ground
(41) ⟶ Tape unload control
(42) ⟶ Tape unload control
(43) ⟶ Capstan Fast forward, rewind & still control
(44) ⟶ Capstan drive control
(45) ⟶ Capstan forward & reverse control
(46) ⟶ Counter control
(47) ⟵ Strobe pulse
(48) ⟷ Data
(49) ⟷ Data
(50) ⟷ Data
(51) ⟷ Data
(52) ⟵ 5v Supply

M50760-365

Audio controller

Package type B (20 pins)

Pins		**Function**
(1) 0v Ground		
(2)	——→	Video input LED drive, Internal = H, External = L
(3)	——→	LED drive, Normal = H, HIFI = L
(4) 0v Ground		
(5) 0v Ground		
(6)	——→	Left audio select, (Normal) = H, (L, L+R) = L, (R) = H
(7)	——→	Right audio select, (Normal) = H, (L, L+R) = L, (R) = L
(8)	——→	E-E delay = L
(9)	——→	E-E delay = L
(10) 4v5		
(11)	——→	Normal volume reset, Internal = L
(12)	——→	Mute = H
(13)	←——	Power on reset = L
(14) 2v5, 300kHz oscillator		
(15) 2v5, 300kHz oscillator		
(16)	——→	E-E = H, Playback = L
(17)	←——	Left audio select, Normal = H (As pin 6)
(18)	←——	Right audio select, Normal = H (As pin 7)
(19)	←——	Video, internal =H, External = L
(20)	←——	5v Supply

M50957/M50959

Voltage synthesizer tuning

Package type B (64 pins)

Pins		Function
(1)	←	5v Supply
(2)	→	External AV = H
(3)	→	Simulcast on = H
(4)	→	Tuning band select, VHF low = L, VHF high = H, UHF = H
(5)	→	Tuning band select, VHF low = L, VHF high = H, UHF = L
(6)	→	Tuning Pulse width modulated voltage
(7)	→	On screen display, select = L
(8)	→	Clock pulse to memory & On screen display i.c
(9)	↔	Data bus memory & On screen display i.c
(10)	→	Memory i.c enable = L
(11)	→	Transmission system select, Pal = H, Secam = L
(12)	→	Blue background = H
(13)	→	Tuner on = H
(14)	→	Audio mute = L
(15)	→	Auto' fine tuning = H
(16)	→	Memory i.c 2 enable = L
(17)	→	Clock pulse
(18)	→	Data bus
(19)	←	Data bus
(20)	←	Auto' fine tuning curve
(21)	←	Horizontal sync'
(22)	→	Clock pulse
(23)	↔	Data bus
(24)	←	Power failure monitoring
(25)	←	Remote control
(26) 0v Ground		
(27)	←	Reset
(28), (29) Xtal 6.29 MHz		
(30), (31) Xtal 32.768kHz power failure mode		
(32) 0v Ground		
(33)		
(34)	←	Key scan
(35)	←	Key scan
(36)	←	Key scan
(37)	←	Key scan

(Continued opposite)

(38) –ve30v
(39) ⟶ Stereo Led = H
(40) ⟶ Stereo Led = H
(41) ⟶ Display segment & Key scan
(42) ⟶ Display segment & Key scan
(43) ⟶ Display segment & Key scan
(44) ⟶ Display segment & Key scan
(45) ⟶ Display segment & Key scan
(46) ⟶ Display segment & Key scan
(47) ⟶ Display segment & Key scan
(48) ⟶ Display grids
(49) ⟶ Display grids
(50) ⟶ Display grids
(51) ⟶ Display grids
(52) ⟶ Display grids
(53) ⟶ Display grids
(54) ⟶ Display grids
(55) ⟶ Display grids
(56) ⟶ Display grids
(57) ⟶ Display segments & Key scan
(58) ⟶ Display segments & Key scan
(59) ⟶ Display segments & Key scan
(60) ⟶ Display segments & Key scan
(61) ⟶ Display segments & Key scan
(62) ⟶ Display segments & Key scan
(63) ⟶ Display segments & Key scan
(64) ⟶ Display segments & Key scan

M51015L

Remote control receiver

Package type A (16 pins)

Pins		**Function**
(1)	←	Infra - red diode
(2)		
(3)	0v Ground	
(4)	→	Amplified signal to pin 6
(5)	RC Filter & waveform shaping circuit	
(6)	←	From pin 4
(7)	RC Filter & waveform shaping circuit	
(8)		
(9)	0v Ground	
(10)		
(11)		
(12)		
(13)		
(14)	0v Ground	
(15)	←	9v Supply
(16)	→	Data output

M51292

Audio and Video signal input selection

Package type B (20 pins)

Pins		Function
(1) 0v	←	Audio input 1
(2)	←	Audio input Logic control
(3)	←	Audio input Logic control
(4)	→	To pin 8 scart socket
(5)	←	12v Supply
(6)	←	Audio input 2
(7)	←	Audio input 3
(8) 0v Ground		
(9)	←	Mono audio input
(10)	→	Audio out from either pin 1,6 or 7
(11) 0v Ground		
(12)	←	Video input 1
(13) 0v Ground		
(14)	←	Video input 2
(15)	←	Video input logic control
(16)		
(17)	←	Video input logic control
(18)	←	Video input 3
(19)	←	12v Supply
(20)	→	Video output from either pin 12,14 or 18

M51293

Dual 5 × single pole, to 1 output audio & video selection switch

Package type B (24 pins)

Pins		Function
(1)	←	See note
(2)	←	Switch select Logic control
(3)	→	See note
(4)	0v Ground	
(5)	←	Pole 0 Switch A
(6)	0v Ground	
(7)	←	Pole 1 Switch A
(8)	←	Pole 2 Switch A
(9)	←	Pole 3 Switch A
(10)	←	Pole 4 Switch A
(11)	←	–ve 6v Supply
(12)	→	Switch A selected output
(13)	→	Switch B selected output
(14)	←	+ve 6v Supply
(15)	←	Pole 0 Switch B
(16)	←	Pole 1 Switch B
(17)	←	Pole 2 Switch B
(18)	←	Pole 3 Switch B
(19)	0v Ground	
(20)	←	Pole 4 Switch B
(21)	←	Switch select logic control
(22)		See note
(23)	Switch select logic control	
(24)	See note	

Signal entering pin1 passes through a 14db amplifier and exits at pin 3.
Likewise a signal entering at pin 24 and exits at pin 22
Switch designated letters, i.e A, B are only given to distinguish between the individual devices
For clarity the signal is shown entering the switches via their pole terminals. In practice the signal direction may be the reverse.

M51316P

Vision & sound I.F detector

Package type B (30 pins)

Pins		Function
(1) 4v2	←	RF automatic gain control
(2) 4v8		
(3) 4v8		
(4)	←	12v Supply
(5) 0v5		
(6) 0v Ground		
(7) 4v1		
(8) 4v1	←	I.F from SAW filter
(9) 4v1	←	I.F from SAW filter
(10) 4v1		
(11) 0v Ground		
(12)		
(13) 4v7		
(14) 7v0, RC Sound FM detector circuit		
(15) 5v5, 5.5 MHz xtal sound detector circuit		
(16) 6v2		
(17) 4v0	→	Audio output
(18) 4v6	←	Sound IF from pin 20 via 5.5MHz filter
(19) 4v6		
(20) 4v6	→	To pin 18
(21) 6v6		
(22) 2v8		
(23) 2v8		
(24) 5v4	→	Automatic frequency control
(25) 3v5		
(26) 6v4, LC Vision detector circuit		
(27) 6v4, LC Vision detector circuit		
(28) 3v5		
(29) 7v1	→	Video output
(30) 6v1	→	Tuner automatic gain control

M51362SP

Sound & vision I.F detection

Package type B (30 pins)

Pins		Function
(1)	→	AGC to tuner
(2)		
(3)	→	Video output
(4)	←	Tuner AGC adjustment
(5)	←	9v Supply
(6) I.F agc		
(7) 0v Ground		
(8)	←	Vision I.F from SAW Filter
(9)	←	Vision I.F from SAW Filter
(10) 0v Ground		
(11)	←	Sound I.F from SAW Filter
(12)	←	9v Supply
(13)		
(14)		
(15)	→	Sound I.F output
(16)	→	Tuner AFC control
(17) FM detector LC circuit		
(18) FM detector LC circuit		
(19)	→	Audio output
(20)		
(21)	←	9v Supply
(22) Detectors VCO (voltage controlled oscillator). LC circuit		
(23) Detectors VCO (voltage controlled oscillator). LC circuit		
(24) 0v Ground		
(25)		
(26) AFC, LC circuit		
(27) AFC, LC circuit		
(28)	→	Demodulated video to pin 30
(29)		
(30)	←	From pin 28

M51365SP

Sound & vision I.F detector

Package type B (30 pins)

Pins		Function
(1) 4v0	⟶	Tuner AGC control
(2) 4v0		
(3) 5v0	⟶	Video output
(4) 3v3	⟵	AGC control
(5)	⟵	9v Supply
(6) 8v6		
(7) 0v Ground		
(8) 2v3	⟵	I.F from SAW Filter
(9) 2v3	⟵	I.F from SAW Filter
(10) 0v Ground		
(11) 4v7		
(12)	⟵	12v Supply
(13) 2v7	⟵	6MHz sound from pin 28
(14) 2v7		
(15) 3v7		
(16) 0v4		
(17) 5v2	⟵	Audio from pin 18 via 6MHz filter
(18) 4v5	⟶	Sound to pin 17
(19) 6v3	⟶	Audio output
(20) 4v0		
(21)	⟵	9v Supply
(22) 8v0, Vision detector circuit		
(23) 8v0, Vision detector circuit		
(24) 0v Ground		
(25) 4v6		
(26) 3v5		
(27) 3v5	⟶	Automatic frequency control
(28) 3v5	⟶	Detected Video to pin 30 & Sound I.F to FM detector circuit
(29) 0v		
(30) 3v5	⟵	Video from pin 28 via 6MHz sound trap

M51366SP

Sound & vision I.F detector

Package type B (30 pins)

Pins		Function
(1) 8v0	⟶	AGC to tuner
(2) 4v4		
(3) 4v6	⟶	Video out
(4) 3v2	⟵	RF AGC control
(5)	⟵	9v Supply
(6) 6v3	⟵	Mute control
(7) 6v2		
(8) 2v4	⟵	Vision I.F from SAW filter
(9) 0v	⟵	Vision I.F from SAW filter
(10) 0v Ground		
(11) 4v7	⟵	Sound I.F from SAW filter
(12)	⟵	9v Supply
(13)	⟵	Sound I.F from pin 15 via 5.5MHz band pass filter
(14) 0v5		
(15)	⟶	Sound I.F to pin 13
(16) 6v4	⟶	AFC out
(17) 6v0, FM detector LC circuit		
(18) 4v1, FM detector LC circuit		
(19) 4v4	⟶	Audio output
(20) 4v3		
(21)	⟵	9v Supply
(22) 8v2		
(23) 8v2		
(24) 0v Ground		
(25) 5v0		
(26) 3v5, AFC detector LC circuit		
(27) 3v5, AFC detector LC circuit		
(28) 3v0	⟶	Video to pin 30
(29) 0v2		
(30) 3v6	⟵	Video from pin 28

M51496P

Vision & sound I.F amplifier

Package type B (18 pins)

Pins		Function
(1) 4v4, I.F, AGC filter		
(2) 4v4, I.F, AGC filter		
(3) 2v2	←	RF, AGC control
(4) 2v5	←	I.F from SAW filter
(5) 2v5	←	I.F from SAW filter
(6) 0v Ground		
(7)	←	9v Supply
(8)	←	Audio volume control
(9) 4v5	→	1v4 pk-pk Audio output
(10) FM detector LC circuit		
(11)	→	De emphasis audio out
(12) Auto fine tuning CL circuit		
(13)	→	Automatic fine tuning
(14) 4v7, Video detector LC circuit		
(15) 4v7, Video detector LC circuit		
(16) 2v7	←	200mV pk-pk Sound I.F
(17) 4v5	→	Auto gain control to Tuner
(18) 3v1	→	2v2 pk-pk Composite Video out & to pin 16 via sound bandpass filter

Pin 8 may be connected to +ve supply
Pin 11 may be connected to ground via a capacitor

M51712FP

Drum motor drive

Package type B (36 pins)

Pins		**Function**
(1)	←	12v Supply
(2)		
(3)	→	+ve Hall effect device supply voltage
(4)	→	–ve Hall effect device supply voltage
(5)		
(6) 0v Ground		
(7) 0v Ground		
(8)		
(9)		
(10)		
(11)		
(12) 0v Ground		
(13)		
(14)	←	Drum motor control from servo
(15)		
(16)		
(17)		
(18)		
(19) 0v Ground		
(20)		
(21)	←	+Hall effect device 2
(22)	←	– Hall effect device 2
(23)	←	– Hall effect device 3
(24)	←	+ Hall effect device 3
(25) Linked to pin 30		
(26)		
(27)		
(28)		
(29)		
(30) Linked to pin 25		
(31)	←	– Hall effect device 1
(32)	←	+ Hall effect device 1
(33)		
(34)	→	Stator coil 2 drive
(35)	→	Stator coil 3 drive
(36)	→	Stator coil 1 drive

M51722P

Capstan servo discriminator

Package type B (16 pins)

Pins		Function
(1) 0v5	→	5v0 pk-pk frequency doubled FG pulse
(2) 4v3	→	FG pulse
(3) 4v6	←	FG pulse from pin 5
(4) 4v6		
(5) 4v7	→	2v8 pk-pk FG pulse to pin 3
(6)	←	Capstan motor FG +ve pulse
(7) 4v6	←	Capstan motor FG –ve pulse
(8)	←	9v Supply
(9) 0v Ground		
(10) 0v3	←	8v0 pk-pk counted down FG pulse
(11)	←	Capstan discriminator control
(12) 4v6		
(13) 4v6	→	FG frequency to voltage conversion to pin 14
(14) 4v7	←	From pin 13
(15) 4v6	→	External gain control & motor control circuit
(16) 0v3		

M5172L

Drum motor drive

Package type A (12 pins)

Pins		Function
(1)	←	Drive from servo
(2) n/c		
(3)	←	Hall effect device 2
(4)	←	Hall effect device 1
(5)	←	Common return from Hall effect device
(6)	→	Hall effect device bias
(7)	←	13v Supply
(8)	→	+ve Stator coil drive 1
(9)	→	–ve Stator coil drive 1
(10)	←	13v Supply
(11)	→	–ve Stator coil drive 2
(12)	→	+ve Stator coil drive 2

M51796P

ACE control head & Capstan FG Amplifier

Package type B (14 pins)

Pins		Function
(1) 0v Ground		
(2) 2v0		
(3)	←———	Control track pulse from ACE head
(4) 2v5		
(5)	———→	Amplified & squared control pulses to pin 6
(6)	←———	5v0 pk-pk control pulses to from pin 5
(7)	———→	Control pulses to Servo circuit
(8)	←———	5v Supply
(9) 2v5	———→	5v0 pk-pk Capstan FG pulses to Servo circuit
(10) 2v6		
(11)	←———	Capstan FG pulses from pin 12
(12)	———→	Amplified Capstan FG pulses to pin 11
(13)	←———	Capstan FG pulses
(14) 0v Ground		

M5201L

2 input amplifiers to 1 switched output — Package type A (8 pins)

Pins		Function
(1)	←	Amplifier 1 or 2, output select
(2) 5v7	←	Amplifier 1, non inverted input
(3) 5v7	←	Amplifier 1, inverted input
(4) 0v Ground		
(5) 5v8	→	Amplifier 1 or 2 output
(6) 5v7	←	Amplifier 2, inverted input
(7) 5v7	←	Amplifier 2, non inverted input
(8)	←	12v Supply

Amplifiers designated numbers, i.e 1,2 are only given to distinguish between the individual devices

M52055

3 × single pole, 2 way switches

Package type B (16 pins)

Pins		**Function**
(1)	⟶	Switch 1, terminal 1
(2)	⟵	Switch 1 pole position control,
(3)	⟵	Switch 1 pole terminal
(4) 0v Ground		
(5)	⟵	Switch 2 pole terminal
(6)	⟵	Switch 3 pole terminal
(7)	⟵	Switch 3, pole position control
(8)	⟶	Switch 3, terminal 1
(9)	⟶	Switch 3, terminal 2
(10) 0v Ground		
(11)	⟶	Switch 2, terminal 2
(12)	⟵	Switch 2, pole position control
(13)	⟵	5v Supply
(14)	⟶	Switch 2, terminal 1
(15) 0v Ground		
(16)	⟶	Switch 1, terminal 2

Switch designated numbers, i.e 1,2, etc are only given to distinguish between the individual devices
For use in video & audio signal circuits.
For clarity the signal route is shown entering the pole terminals at pins 3, 5 & 6. Its exit, depending on the control logic inputs, will be from one of the remaining terminals. Its helpful to visualize the device as a mechanical 3 × rotary type wafer switch. i.e, single pole × 2 way.

M52057

Chroma processor

Package type C (48 pins)

Pins		Function
(1) n/c		
(2)	←	520mV pk-pk, 5.06MHz from sub convertor pin 36
(3) 2v0	←	4v4 pk-pk, 25Hz flip flop
(4) 3v3		
(5) 2v4	→	Playback mode 1v8 pk-pk, 4.43MHz chroma to pin 17 Record mode 500mV pk-pk, video to pin 17
(6) 2v8	→	Record mode 550mV pk-pk, 627kHz chroma to Secam detector Playback mode 1v6 pk-pk, 627kHz chroma to Secam detector
(7)	←	5v Supply
(8) 2v8	←	Record mode 520mV pk-pk, video input
(9) 3v4		
(10) 2v3	→	Playback mode 400mV pk-pk, FM video to pin 15 Record mode 1v0 pk-pk, 627kHz chroma to pin12
(11) 3v0	←	Playback mode 180mV pk-pk, FM video input
(12) 3v0	←	Record mode 320mV pk-pk, from pin 10
(13) 2v4	→	Record mode 480mV pk-pk, 627kHz output
(14) n/c		
(15) 3v0	←	Playback mode 150mV pk-pk, 627kHz chroma from pin 10
(16) 2v9		
(17) 2v9	←	From pin 5
(18)	←	Pal = H, Secam = L
(19) 0v Ground		
(20) 3v4		
(21)	←	Trick play modes control
(22) 3v3		
(23) 3v0		
(24) 2v6		
(25) n/c		
(26) 3v3		
(27) 3v3		
(28) 3v8		
(29)	→	110mV pk-pk, 4.43MHz oscillator output
(30) 3v0, 1v0 pk-pk. 4.43MHz xtal		
(31)	←	5v Supply

(Continued opposite)

(32) 3v6,1v0 pk-pk. 4.43MHz xtal
(33) 0v6
(34) 0v3 ←——— 4v4 pk-pk, Horizontal sync' pulse
(35) 2v7
(36) 2v5 1v0 pk-pk, 5.06Mhz sub convertor to pin 2
(37) 0v4
(38) n/c
(39) 4v6
(40) ←——— LP = H, SP = L
(41) ←——— Playback = H, Record = L
(42) ———→ Playback, 850mV pk-pk, 4.43Mhz chroma output
(43) n/c
(44) 0v Ground
(45) 3v0 ←——— Playback 110mV pk-pk, 4.43MHz chroma from 2 line delay
(46) 3v2
(47) 3v2 ———→ Playback 4.43MHz chroma to pin 45 via 2 line delay
(48) 3v4

M5218LO

Dual operational amplifier

Package type A & B (8 pins)

Pins		Function
(1)	⟶	Amplifier 1 output
(2)	⟵	Amplifier 1 Inverted input
(3)	⟵	Amplifier 1 non inverted input
(4)	0v Ground	
(5)	⟵	Amplifier 2 non inverted input
(6)	⟵	Amplifier 2 inverted input
(7)	⟶	Amplifier 2 output
(8)	⟵	12v Supply

Also used as headphone output
Amplifiers designated numbers, i.e 1,2 are only given to distinguish between the individual devices

M5238

Dual operational amplifier

Package type B
(8 pins)

Pins		Function
(1)	⟶	Amplifier 1 output
(2)	⟵	Amplifier 1 non inverted input
(3)	⟵	Amplifier 1 inverted input
(4)	0v Ground	
(5)	⟵	Amplifier 2 non inverted input
(6)	⟵	Amplifier 2 inverted input
(7)	⟶	Amplifier 2 output
(8)	⟵	12v Supply

M5241L

2 channel (stereo) audio amplifier

Package type A (10 pins)

Pins		Function
(1) 4v8	→	Amplifier 1 output
(2) 5v4	←	Amplifier 1 input
(3) 4v8		
(4) 0v Ground		
(5) 4v8		
(6)	←	Amplifier 1 volume control
(7)	←	Amplifier 2 volume control
(8)	←	Amplifier 2 input
(9)	→	Amplifier 2 output
(10)	←	12v Supply

Amplifier designated numbers, i.e 1,2,are only given to distinguish between the individual devices

M54543L

Loading motor drive

Package type A
(9 pins)

Pins		Function
(1)	←	12v Supply
(2)	←	12v Supply
(3)	→	Loading motor drive
(4)	←	Loading control
(5) 0v Ground		
(6)	←	Unloading control
(7)	→	Loading motor drive
(8)	←	12v Supply
(9)	←	12v Supply

Also used for reel motor applications

M54548L

Capstan motor drive

Package type A (12 pins)

Pins			Function
(1)		→	Motor drive forward
(2)	0v Ground		
(3)		→	Motor drive reverse
(4)	5v1		
(5)	8v3		
(6)		←	12v Supply
(7)	3v8		
(8)	2v9	←	Drive from servo
(9)	10v5		
(10)		←	Forward, Reverse, stop & brake Capstan mode control
(11)		←	Forward, Reverse, stop & brake Capstan mode control
(12)		←	Forward, Reverse, stop & brake Capstan mode control

M54549L

Motor driver

Package type A (12 pins)

Pins		Function
(1)	⟶	Driver 2 output to carriage motor +ve terminal
(2)	⟵	10v Supply
(3)	⟵	Motor 12v Supply
(4)	⟶	Driver 2 output to carriage motor –ve terminal
(5) 0v Ground		
(6)	⟵	Hi = Driver 2, Lo = Driver 1 select control
(7)	⟵	Logic control to selected driver
(8)	⟵	Logic control to selected driver
(9)	⟶	Driver 1 output to loading motor +ve terminal
(10)	⟵	Motor 12v Supply
(11)	⟵	10v Supply
(12)	⟶	Driver 1 output to loading motor –ve terminal

This device contains two motor drivers. For clarity designated driver 1 & driver 2 The example shows Driver 1 connected to a loading motor and driver 2 connected to a carriage up / down motor.

M54644BL

Reel motor control

Package type A (9 pins)

Pins		**Function**
(1) 0v		
(2)	←	Reel motor supply voltage control
(3)	→	+ve Reel motor drive
(4)	←	–ve Reel motor control
(5) 0v Ground		
(6)	←	+ve Reel motor control
(7)	→	–ve Reel motor drive
(8)	←	Reel motor supply voltage control
(9)	←	12v Supply

Capstan motor control

(1)		
(2)	←	17v Supply
(3)	→	+ve Capstan motor drive
(4)	←	–ve Capstan motor control
(5) 0v Ground		
(6)	←	+ve Capstan motor control
(7)	→	–ve Capstan motor drive
(8)	→	Capstan motor supply voltage control i.e pause.
(9)	←	12v Supply

M54647L

Reel motor control

Package type A (12 pins)

Pins		Function
(1) 0v		
(2)	←	12v Supply
(3)	←	Motor 12v Supply
(4)	→	Motor drive, Forward = L, Reverse = H, Brake = L, Stop = Open
(5)	←	Motor control, Forward = L, Reverse = H Brake = H, Stop = L
(6) 0v Ground		
(7) 0v Ground		
(8)	←	Motor control, Forward = H, Reverse = L Brake = H, Stop = L
(9)	→	Motor drive, Forward = H, Reverse = L Brake = L, Stop = Open
(10)		
(11)	←	Additional control from system control circuit
(12)	←	Additional Pulse width modulated control from system control circuit

M54648L

Capstan motor drive

Package type A (12 pins)

Pins		Function
(1)	——→	Forward motor drive
(2) 0v Ground		
(3)	——→	Reverse motor drive
(4) Slow mode circuit		
(5) Fast forward & rewind circuit		
(6)	←——	12v Supply
(7)		
(8)	←——	Servo drive
(9)	←——	Pause control
(10)	←——	Motor modes control
(11)	←——	Motor modes control
(12)	←——	Motor modes control

M54649L

3 × Motor drive control

Package type A (10 pins)

Pins		Function
(1)	0v Ground	
(2)	0v5 ⟶	Unload drive to loading motor
(3)	0v5 ⟶	Unload drive to carriage motor
(4)	⟵	Mode instructions from system control
(5)	⟵	Mode instructions from system control
(6)	⟵	Mode instructions from system control
(7)	⟵	12v Supply
(8)	8v0	
(9)	⟵	12v motors supply
(10)	0v5 ⟶	Load drive to loading & carriage motors

Example shows the device used as a drive for a carriage & loading motor combination

M54680

Capstan motor drive

Package type C (42 pins)

Pins		Function
(1)	←	+ve Hall effect device 1
(2)	←	–ve Hall effect device 1
(3)	←	+ve Hall effect device 2
(4)	←	–ve Hall effect device 2
(5)	←	+ve Hall effect device 3
(6)	←	–ve Hall effect device 3
(7)		
(8) n/c		
(9) 0v Ground		
(10) 0v Ground		
(11) 0v Ground		
(12) 0v Ground		
(13) 0v Ground		
(14)	←	Output current sensing
(15)	←	Search speed control
(16)		
(17)		
(18)	←	FG pulses
(19)	←	FG pulses
(20)		
(21)	←	Shaped FG pulses out
(22) n/c		
(23) n/c		
(24)	←	Reference voltage
(25) n/c		
(26)		
(27)	←	Servo drive
(28)	←	Speed high / low control
(29)	←	5v Supply
(30) 16v0		
(31)	←	16v Supply
(32)		
(33) 0v Ground		
(34) 0v Ground		
(35) 0v Ground		
(36) 0v Ground		
(37) 0v Ground		
(38) n/c		
(39) 14v0		
(40)	→	Drive to stator coil 1
(41)	→	Drive to stator coil 2
(42)	→	Drive to stator coil 3

M54898P

Capstan & Drum servo

Package type B (64 pins)

Pins		Function
(1)	←	5v Supply
(2)	←	3v0 pk-pk ramp waveform from manual tracking control
(3) 0v5 pk-pk sawtooth waveform		
(4)	←	2v0 pk-pk tach pulse from Drum motor
(5)	←	2v0 pk-pk tach pulse from Drum motor
(6) n/c		
(7)		
(8) n/c		
(9) 5v0	→	SP speed detection
(10)		
(11)	→	5v0 pk-pk squarewave Capstan phase control to pin 25
(12) 1v0	←	search mode control
(13)	→	Pause playback mode control
(14)		
(15) n/c		
(16)	→	Capstan motor slow mode control
(17)	←	Fine still mode control
(18)	←	Fine still mode control
(19)	←	Frame advance control
(20)		
(21)		
(22) 2v6	→	5v0 pk-pk squarewave Capstan speed control to pin 24
(23)	←	Pause, frame advance control
(24)	←	From pin 22
(25)	←	From pin 11
(26)	←	Playback pause control
(27) 2v6	→	Capstan drive
(28)	←	–ve 5v0 pk-pk Capstan FG pulses
(29)	←	+ve 5v0 pk-pk Capstan FG pulses
(30)	→	5v0 pk-pk Capstan FG pulses to system control circuit
(31) 2v5		
(32) 2v4	←	250mV pk-pk, 4.43MHz clock pulse
(33)		
(34)		
(35) 2v4		
(36) 2v0		

(Continued overleaf)

M54898P

(37) 2v5
(38) 4v5
(39) 2v7 ⟶ Drum motor circuit drive
(40) ⟵ Drum speed control from pin 43
(41) 2v5 ⟵ Drum phase control from pin 42
(42) ⟶ 5v0 pk-pk squarewave drum phase control to pin 41
(43) ⟶ 5v0 pk-pk squarewave drum speed control to pin 40
(44) ⟶ 25Hz flip flop pulse
(45) ⟵ Head switching point adjustment
(46) 2v7 pk-pk vertical sync' pulse
(47) 0v5 ⟵ VStill mode vertical lock adjustment
(48)
(49) ⟵ Picture reverse mode control
(50) ⟵ 3v0 pk-pk drum FG pulses
(51) 800mV pk-pk sawtooth waveform
(52) 400mv pk-pk squarewave
(53) ⟵ Record mode 4v0 pk-pk vertical sync' pulse
(54) ⟵ Record mode = 2v0, Playback = 0v
(55) ⟶ Luminance & chroma mute control
(56) n/c ⟶ Drum FG pulses
(57) n/c
(58) n/c
(59) ⟶ Record mode 3v0 pk-pk tracking control pulse to ACE head
⟵ Playback tracking control pulse from ACE head
(60) ⟵ ACE head common
(61) 3v5 pk-pk playback control pulse
(62) ⟶ Amplified 3v0 pk-pk playback tracking control pulse to pin 63
(63) ⟵ From pin 62
(64) 0v Ground

M58655P

Memory

Package type B
(14 pins)

Pins		Function
(1)	←—	5v Supply
(2)	←—	–ve 30v Supply
(3) n/c		
(4)	←—	Chip select, data transfer = L
(5) n/c		
(6)	←—	Clock pulse
(7)	←—	Read, write, erase
(8)	←—	Read, write, erase
(9)	←—	Read, write, erase
(10) n/c		
(11) n/c		
(12)	←→	Data
(13) 0v Ground		
(14) n/c		

M58659P

Memory

Package type B
(14 pins)

Pins		**Function**
(1)	←——	5v Supply
(2)	←——	–ve 30v Supply
(3) n/c		
(4)	←——	Chip select command
(5) n/c		
(6)	←——	Clock pulse
(7)	←——→	Read, write, erase data
(8)	←——→	Read, write, erase data
(9)	←——→	Read, write, erase data
(10) n/c		
(11) n/c		
(12)	←——	Memory data
(13) n/c		
(14) n/c		

Reading & writing polarities
Pin 7
S/By = H, Erase = H, Address input = H, Read = L, Data out = L, Write = L, Data in = L
Pin 8
S/by = H, Erase = L, Address input = L, Read = H, Data out = H, Write = L, Data in = L
Pin 9
S/by = H, Erase = H, Address input = L, Read = H, Data out = L, Write = H, Data in = L

Capstan motor control

Package type B (16 pins)

Pins		Function
(1) 12v	⟶	Capstan motor control circuit voltage
(2)	⟶	Normal speed = H, Still mode = L
(3)	⟵	Still mode pulse command
(4)	⟵	5v0 pk-pk Drum FG pulse
(5)	⟵	Motor stop = H, Motor go = L
(6) Clock RC circuit		
(7) Clock RC circuit		
(8) 0v Ground		
(9)	⟵	Head switching pulse
(10)	⟵	ACE Control pulse
(11)	⟵	Drop out compensation pulse from head preamplifier circuit
(12)	⟵	Motor reverse mode control Motor reverse from forward = L, Motor reverse from reverse = H
(13)	⟶	Capstan motor control, Rewind = H, Forward = L, Stop = L
(14)	⟶	Capstan motor control, Rewind = L, Forward = H, Stop = H
(15)	⟵	Power reset = L
(16)	⟵	5v Supply

MB88501-147M

System control processor | **Package type B (42 pins)**

Pins		Function
(1)	↔	Data, expander & memory i.c's
(2)	↔	Data, expander & memory i.c's
(3)	↔	Data, expander & memory i.c's
(4)	↔	Data, expander & memory i.c's
(5)	→	Audio playback mode = L
(6)	→	Audio record mode = L
(7)	→	Audio dub = L
(8)	→	Tape speed slow = L
(9)	→	Still mode = L
(10)	↔	Data, external expander i.c timing circuits
(11)	↔	Data, external expander i.c timing circuits
(12)	←	Take up reel rotation pulse
(13)	←	Head switching pulse
(14)	←	Video normal = H
(15)	←	Auto fine tuning normal = H
(16) 4MHz xtal		
(17) 4MHz xtal		
(18)	←	Power on reset = L
(19)	←	Trigger pulse for dummy vertical sync' in trick modes
(20)	←	Power failure detection, memory save
(21) 0v Ground		
(22)	→	Clock pulse, operations i.c
(23)	↔	Data operation i.c
(24)	↔	Data operation i.c
(25)	→	Address external memory
(26)	→	Address external memory
(27)	→	Address external memory
(28)	→	Address external memory
(29)	→	Address external memory
(30)	→	Address external memory
(31)	→	Address external memory
(32)	→	Address external memory
(33)	→	Address external memory
(34)	→	Address external memory
(35)	→	Read enable to external memory i.c
(36)	→	Write enable to external memory i.c
(37)	←	5v Supply
(38)	→	Reset to external i.c = L
(39)	→	Vertical sync' pulse for trick modes
(40)	→	Data
(41) 0v Ground		
(42)	←	5v Supply

MC14051

8 position single pole logic controlled switch

Package type B (16 pins)

Pins		Function
(1)	→	Terminal 2
(2)	→	Terminal 1
(3)	←	Pole terminal
(4)	→	Terminal 8
(5)	→	Terminal 7
(6)	0v Ground	
(7)	←	–ve 5v Supply
(8)	0v Ground	
(9)	←	Control logic
(10)	←	Control logic
(11)	←	Control logic
(12)	→	Terminal 6
(13)	→	Terminal 5
(14)	→	Terminal 4
(15)	→	Terminal 3
(16)	←	+ve 5v Supply

For use in video & audio signal circuits.
For clarity the signal route is shown entering the pole terminal at pin 3. Its exit, depending on the control logic inputs, will be from one of the remaining terminals. It is helpful to visualize the device as a mechanical rotary type wafer switch. i.e, 1 pole × 8 way.

MC14052

Dual 4 position single pole switch | **Package type B (16 pins)**

Pins		Function
(1)	⟶	Switch 1, terminal 2
(2)	⟶	Switch 1, terminal 1
(3)	⟵	Switch 1 pole terminal
(4)	⟶	Switch 1, terminal 4
(5)	⟶	Switch 1, terminal 3
(6) 0v Ground		
(7)	⟵	–ve 5v Supply
(8) 0v Ground		
(9)	⟵	Control Logic
(10)	⟵	Control Logic
(11)	⟶	Switch 2, terminal 2
(12)	⟶	Switch 2, terminal 1
(13)	⟵	Switch 2 pole terminal
(14)	⟶	Switch 2, terminal 4
(15)	⟶	Switch 2, terminal 3
(16)	⟵	+ve 5v Supply

For use in video & audio signal circuits.
For clarity the signal route is shown entering the pole terminals at pins 3 & 13. Its exit, depending on the control logic inputs, will be from one of the remaining terminals. Its helpful to visualize the device as a dual mechanical rotary type wafer switch. i.e, 1 pole × 4 way.

MC14053

3 × single pole 2 way switches

Package type B (16 pins)

Pins		**Function**
(1)	——→	Switch 1, terminal 2
(2)	——→	Switch 1, terminal 1
(3)	——→	Switch 2, terminal 1
(4)	←——	Switch 2, pole terminal
(5)	——→	Switch 2, terminal 2
(6)	0v Ground	
(7)	←——	–ve 5v Supply
(8)	0v Ground	
(9)	←——	Control logic
(10)	←——	Control logic
(11)	←——	Control logic
(12)	——→	Switch 3, terminal 2
(13)	——→	Switch 3, terminal 1
(14)	←——	Switch 3, pole terminal
(15)	←——	Switch 1, pole terminal
(16)	←——	+ve 5v Supply

For use in video & audio signal circuits.
For clarity the signal route is shown entering the pole terminals at pins 4, 14 & 15. Its exit, depending on the control logic inputs, will be from one of the remaining terminals. Its helpful to visualize the device as a mechanical 3 × rotary type wafer switch. i.e, single pole × 2 way.

MC14519

4 × single pole 2 way switches

Package type B (16 pins)

Pins		Function
(1)	⟶	Switch 1, terminal 1
(2)	⟶	Switch 2, terminal 1
(3)	⟶	Switch 2, terminal 2
(4)	⟶	Switch 3, terminal 1
(5)	⟶	Switch 3, terminal 2
(6)	⟶	Switch 4, terminal 1
(7)	⟶	Switch 4, terminal 2
(8) 0v Ground		
(9)	⟵	H = Control logic, all switch poles make to terminals 1 L = Control logic, all switch poles make to terminals 2
(10)	⟵	Switch 4, pole terminal
(11)	⟵	Switch 3, pole terminal
(12)	⟵	Switch 2, pole terminal
(13)	⟵	Switch 1, pole terminal
(14)	⟵	H = Control logic, all switch poles make to terminals 2 L = Control logic, all switch poles make to terminals 1
(15)	⟶	Switch 1, terminal 2
(16)	⟵	5v Supply

Switch designated numbers, i.e 1,2, etc are only given to distinguish between the individual devices.
For clarity the signal routes are shown entering the pole terminal at pins 10,11,12 & 13. they exit, depending on the control logic inputs, from one of the remaining terminals.

MN15311

Slow & fine speed control servo

Package type B (28 pins)

Pins		Direction	Function
(1)		←	5v Supply
(2)		←	Serial data
(3)		←	Serial clock pulse
(4)	2v5	←	25Hz Head switching pulse
(5)			
(6)	2v5	←	Capstan drive sample
(7)		→	Current limiting to capstan drive
(8)		←	From pin 12
(9)	n/c		
(10)		→	Capstan drive control, Reverse = H, Stop = Medium, Forward = L
(11)		←	SP mode = L
(12)		→	Capstan brake control to pin 8
(13)		→	Head amplifier 25Hz switching pulse
(14)	4v5	←	FM envelope detection
(15)	2v5	→	5v0 pk-pk Flip flop to chroma circuit
(16)		←	Drum motor drive sample
(17)		←	Artificial vertical sync' pulse
(18)		←	Still mode, 5v0 pk-pk vertical sync'
(19)		→	ACE head control pulse
(20)	0v2	←	Slow tracking control
(21)			
(22)		←	4.43MHz clock pulse
(23)	4v7	←	Capstan on / off / reset control
(24)	0v Ground		
(25)			
(26)		←	Capstan FG pulse
(27)		→	Control pulse phase
(28)	2v5		

MN15311

Fine speed & still servo control

Package type B (28 pins)

Pins			**Function**
(1)		←	5v Supply
(2)		←	Serial data
(3)		←	Serial clock pulse
(4)	2v5	←	25Hz Head switching pulse
(5)			
(6)	2v5	←	Capstan drive sample
(7)		→	Current limiting to capstan drive
(8)		←	From pin 12
(9)	n/c		
(10)		→	Capstan drive control, Reverse = H, Stop = Medium, Forward = L
(11)		←	SP mode = L
(12)		→	Capstan brake control to pin 8
(13)		→	Head amplifier switching pulse
(14)	4v5	←	FM envelope detection
(15)	2v5	→	5v0 pk-pk Flip flop to chroma circuit
(16)	2v6	←	Drum motor drive sample
(17)	0v5	←	Artificial vertical sync' pulse
(18)	0v	←	Still mode, 5v0 pk-pk vertical sync'
(19)	2v5	→	ACE head control pulse
(20)	0v2	←	Slow tracking control
(21)	2v5		
(22)	2v2	←	4.43MHz clock pulse
(23)	4v7	←	Capstan on / off / reset control
(24)	0v Ground		
(25)	2v5		
(26)	2v5	←	Capstan FG pulse
(27)	2v6	→	Control pulse phase
(28)	2v5		

MN6168

Drum & Capstan servo

Package type B (28 pins)

Pins		Function
(1) 0v Ground		
(2) 2v6	→	5v0 pk-pk pulse width modulated capstan speed
(3) 3v0	→	5v0 pk-pk pulse width modulated capstan phase
(4)	→	Record mode control pulse to ACE head
(5) 5v0	←	5v0 pk-pk vertical sync' pulse
(6) 2v4	←	Playback tracking control
(7)	←	Record mode pause control = H
(8) 0v Ground		
(9)	←	Record =, Playback = L
(10) 0v Ground		
(11) 5v		
(12) 5v		
(13) 0v Ground		
(14)	←	5v Supply
(15) 1v5	←	4.43MHz clock pulse
(16)	←	5v0 pk-pk vertical sync' capstan phase control
(17) 2v6	←	Capstan phase & speed control from FG pulse circuit
(18)	←	Mode select, capstan reverse = H
(19) 0v Ground		
(20)	←	Mode select, capstan normal speed = L
(21)	←	Mode select, capstan normal speed = H
(22)	←	Mode select, capstan on = H
(23) 5v0		
(24) 4v0	←	Drum FG pulse
(25) 2v5	←	Drum PG pulse
(26) 0v Ground		
(27) 2v8	→	5v0 pk-pk pulse width modulated drum phase control
(28) 2v7	→	5v0 pk-pk pulse width modulated drum speed control

MN675121

Combined servo & system control microcomputer **Package type C (84 pins)**

Pins		Function
(1)	——→	Picture blank control
(2)	——→	Audio playback or record control
(3)	——→	Record mode start control
(4)	——→	Linear audio record start control
(5)	——→	NTSC, PAL system control
(6)		
(7)	←——	Record safety switch
(8)	——→	Serial data
(9)	←——	Supply reel pulses
(10)	←——	Mode sensor
(11)	←——	Mode sensor
(12)	←——	Mode sensor
(13)	←——	Mode sensor
(14)	——→	Loading motor reverse mode
(15)	——→	Loading motor forward mode
(16)	——→	Loading motor torque control
(17)	←——	Serial clock pulses
(18)	←——	Operation circuit micro' serial data
(19)	——→	Operation circuit micro' serial data
(20)	←——	Reset = L
(21) 0v Ground		
(22)	——→	4v0 pk-pk Head switching pulse to head amplifier circuit
(23)	——→	Vertical sync' pulses
(24)	——→	4v0 pk-pk Capstan motor phase control
(25)	——→	4v0 pk-pk Capstan motor speed control
(26)	——→	4v0 pk-pk Drum motor phase control
(27)	——→	4v0 pk-pk Drum motor speed control
(28)	——→	Capstan motor stop = L
(29)	——→	Capstan motor stop control
(30)	——→	Capstan motor gain on = L
(31) System clock xtal		
(32) System clock xtal		
(33) 0v Ground		
(34)	——→	Capstan motor forward or reverse direction control
(35)		
(36)	——→	Audio mute on = L
(37)	——→	Playback mode on = L
(38)	——→	E-E on = L

(Continued opposite)

(39) ——→ PAL = L, MESECAM = H
(40) ——→ Tape trick mode on = H
(41) ——→ Drum motor stop control
(42)
(43) ←—— Head switching pulse control
(44)
(45)
(46) ←—— 1v0 pk-pk Capstan FG pulses
(47) ←—— 100mV pk-pk Drum FG pulses
(48) ←—— 700mV pk-pk Drum PG pulses
(50)
(51)
(52) ←—— Control pulse from pin 62
(53)
(54)
(55) ——→ ACE head record control pulses
(56) ←—— 6v0 pk-pk ACE playback control pulses
(57)
(58)
(59)
(60) 0v Ground
(61) ——→ Control pulse amplifier gain control
(62) ——→ 3v0 pk-pk ACE head control pulse to pin 52
(63) ←—— 5v0 Supply
(64) ←—— 4v0 pk-pk Vertical sync' pulses
(65) ←—— Take up reel pulses
(66) ←—— Video system sensing
(67) ←—— SECAM mode detect
(68) ←—— Video system sensing
(69)
(70) ←—— Video envelope waveform detect
(71) ←—— Tape end sensor
(72) ←—— Tape start sensor
(73) ——→ Picture control
(74) ←—— 5v Supply
(75) ——→ Record timer
(76)
(78) ——→ External input select = L
(79) ——→ NTSC system select
(80) ——→ Tape speed, SP = L
(81) ——→ On screen display i.c, internal or external sync' switch
(82) ——→ Head switching pulse in trick mode
(83) ——→ SP or LP head switching control
(84)

MSM6964

Delay line, Playback drop out compensation

Package type B (8 pins)

Pins			Function
(1)		←	9v Supply
(2)	0v Ground		
(3)		←	5v Supply
(4)	3v5	→	Delayed FM Luminance
(5)	3v1		
(6)	2v4	←	Video
(7)	1v6	←	Clock pulse
(8)	5v		

MSM6965

Luminance Delay line, Playback drop out compensation

Package type B (8 pins)

Pins			Function
(1)		←	9v Supply
(2)	0v Ground		
(3)		←	5v Supply
(4)	3v5	→	Delayed FM Luminance
(5)	3v1		
(6)	2v4	←	FM luminance input
(7)	1v6	←	8.83MHz, 180mV pk-pk Clock pulse
(8)	1v9		

MSM6967

Luminance Delay line, playback drop out compensation — **Package type B (8 pins)**

Pins			Function
(1)		←	12v Supply
(2)	0v Ground		
(3)		←	5v Supply
(4)	3v2	→	Delayed FM luminance output
(5)	3v1		
(6)	2v0	←	Sync' pulses
(7)	1v7	←	FM luminance input
(8)	4v5		

MSM7413

CCD electronic delay line

Package type B
(8 pins)

Pins			**Function**
(1)			
(2)	0v Ground		
(3)		←	5v Supply
(4)	3v5	→	Delayed video output
(5)	1v9		
(6)	2v0	←	Video input
(7)	2v6	←	Clock pulse

NE648N

Dolby audio noise reduction

Package type B (16 pins)

Pins			Function
(1)	5v3	←	Playback audio from pin 7, Record audio from pin 3 via High pass filter
(2)	5v3	←	Record or playback from pin 6 via Low pass filter
(3)	5v2	→	Record mode to pin 1
(4)	5v3		
(5)	5v3	←	Record or Playback mode input
(6)	5v3	→	Amplified pin 5 to pin 2 via Low pass filter
(7)	5v4	→	Record or Playback output
(8)	n/c		
(9)	0v Ground		
(10)	2v1		
(11)	5v4		
(12)	2v1		
(13)	2v1		
(14)		←	Dolby on = L
(15)			
(16)		←	12v Supply

NE648N

Dolby audio noise reduction

Package type B (16 pins)

Pins		Function
(1) 5v3	←	Playback audio from pin 7, Record audio from pin 3 via High pass filter
(2) 5v3	←	Record or playback from pin 6 via Low pass filter
(3) 5v2	→	Record mode to pin 1
(4) 5v4		
(5) 5v3	←	Record or Playback mode input
(6) 5v3	→	Amplified pin 5 to pin 2 via Low pass filter
(7) 5v4	→	Record or Playback output
(8) n/c		
(9) 0v Ground		
(10) 2v2		
(11) 5v4		
(12) 2v2		
(13) 2v2		
(14)	←	Dolby on = L
(15)		
(16)	←	10v Supply

NJM2245

2 Video inputs to 1 output select switch, with a 6db gain amplifier

Package type A (8 pins)

Pins		Function
(1)	←	Video 1input
(2)	←	Select control, Video 1 output = L, Video 2 output = H
(3)	←	Video 2 input
(4) 0v Ground		
(5)		
(6)	←	9v Supply
(7)	→	Amplified output
(8) 0v Ground		

NJM2248

2 × single pole 2 way switches

Package type A (9 pins)

Pins		Function
(1)	——→	Switch 1, terminal A
(2)	←——	9v Supply
(3)	——→	Switch 1, terminal B
(4)	←——	Switch 1, pole terminal
(5) 0v Ground		
(6)	——→	Switch 2, terminal A
(7)	←——	Switch 2 control. H = Switch 2 pole terminal to pin 8 L = Switch 2 pole terminal to pin 6
(8)	——→	Switch 2, terminal B
(9)	←——	Switch 1 control. H = Switch 1 pole terminal to pin 1 L = Switch 1 pole terminal to pin 3

Switch designated numbers, i.e 1,2, are only given to distinguish between the individual devices.
For clarity the signal is shown entering the switches via their pole terminal and exit via terminal A or B. In practice the signal direction could be the reverse.

NJM2901N

4 × operational amplifiers **Package type B (14 pins)**

Pins		**Function**
(1)	⟶	Amplifier 2 output
(2)	⟶	Amplifier 1 output
(3)	⟵	5v Supply
(4)	⟵	Amplifier 1 inverted input
(5)	⟵	Amplifier 1 non inverted input
(6)	⟵	Amplifier 2 inverted input
(7)	⟵	Amplifier 2 non inverted input
(8)	⟵	Amplifier 3 inverted input
(9)	⟵	Amplifier 3 non inverted input
(10)	⟵	Amplifier 4 inverted input
(11)	⟵	Amplifier 4 non inverted input
(12) 0v Ground		
(13)	⟶	Amplifier 4 output
(14)	⟶	Amplifier 3 output

Amplifier designated numbers, i.e 1,2, etc are only given to distinguish between the individual devices

NJM4558

Dual operational amplifier

Package type A
(8 pins)

Pins		Function
(1)	→	Amplifier 1 output
(2)	←	Amplifier 1 Inverted input
(3)	←	Amplifier 1 non inverted input
(4)	0v Ground	
(5)	←	Amplifier 2 non inverted input
(6)	←	Amplifier 2 inverted input
(7)	→	Amplifier 2 output
(8)	←	22v Supply

Alternative supply
Pin 4 –ve 5v Supply
Pin 8 +ve 5v Supply

Amplifiers designated numbers, i.e 1,2,are only given to distinguish between the individual devices

PB20166C

Luminance processor **Package type B (40 pins)**

Pins			Function
(1)	2v6	←	Video from Tuner
(2)		←	5v Supply
(3)	2v7	←	800mV pk-pk video from line input
(4)	3v5		
(5)		→	Playback or record mode video to pin 6 and to chroma circuits
(6)		←	From pin 5
(7)	2v2		
(8)	2v0	→	Composite video to RF Convertor circuit
(9)		←	Vertical pulse for video muting
(10)		→	Playback or record mode, 650mV pk-pk composite video to pin 21 via low pass filter.
(11)			
(12)		←	Playback sharpness control circuit from pin 14
(13)			
(14)		→	To pin 12
(15)		←	550mV pk-pk playback chroma
(16)	0v Ground		
(17)	Playback noise cancelling circuits		
(18)			
(19)			
(20)		→	Sync' pulses to colour & servo circuits
(21)		←	300mV pk-pk Luminance from pin 10
(22)			
(23)	2v5		
(24)			
(25)		←	Playback = H, Record = L
(26)	1v7		
(27)	2v7		
(28)	2v4	→	Luminance to pin 29 via delay line
(29)		←	From pin 28
(30)			
(31)	0v		
(32)		←	Record mode 1v4 pk-pk Luminance from pin 33
(33)		→	To pin 32
(34)			
(35)			

(Continued opposite)

(36) 0v Ground
(37)
(38)
(39) ← Playback FM from video head circuit
(40) → 1v0 pk-pk Record mode FM to video head circuit

PB20227A

Colour processor

Package type B (28 pins)

Pins		Function
(1)	←	5v Supply
(2) 0v		
(3) 2v2		
(4) 1v2		
(5)	←	Video sync' pulses
(6)	←	Tuner = L, Playback = H
(7)	←	Drum flip flop pulse
(8) 3v4	→	Record mode chroma to Luminance FM circuit
(9)	←	Playback chroma from pin 11 via delay line
(10) 2v1		
(11)	→	To pin 9
(12) n/c		
(13)	←	1v0 pk-pk Record mode Video input
(14) SP = L		
(15)	←	200mV pk-pk Chroma from pin 17 via bandpass filter
(16)		
(17)	→	To pin 15
(18)	→	627kHz record chroma to video head circuit
(19)	←	Playback 627kHz chroma from video head circuit
(20)	←	5.06MHz from pin 21
(21)	→	Sub convertor 5.06MHz to pin 20
(22) 2v0		
(23) 2v7		
(24) 0v Ground		
(25)	→	Clock pulse to servo & DOC circuit
(26) 3v6		
(27) 4.43MHz clock xtal		
(28) 4.43MHz clock xtal		

PJ22031A

Luminance playback & record processor

Package type B (44 pins)

Pins		Function
(1) 2v6		
(2) 0v4	→	Record mode Video sync' pulses
(3) 2v8	←	1v pk-pk Composite video
(4) 2v2		
(5) 3v4	←	Playback luminance from pin 24
(6) 1v5		
(7) 1v6	→	Playback luminance to pin 35 via Low pass filter
(8) 0v Ground		
(9)	→	Playback & E-E video to RF convertor
(10)	←	400mV pk-pk Playback colour & Playback / E-E switching
(11)		
(12)	←	Playback luminance from pin 34 via high pass filter
(13)		
(14)	←	Vertical sync' pulse
(15)	→	DOC FM to pin 30
(16)	→	500mV pk-pk demodulated playback FM to pin 29
(17)		
(18) 0v Ground		
(19)	←	5v Playback Supply
(20)	←	Playback FM from DOC delay line
(21)	→	Normal or DOC Playback FM to delay line
(22)	←	300mV pk-pk Playback FM from video head circuit
(23)		
(24)	→	250mV pk-pk Playback luminance from internal sharpness circuit to pin 5
(25)	←	Sharpness control
(26)		
(27)	←	Drum flip flop pulse
(28)	→	Playback video to pin 31
(29)	←	From pin 16
(30)		
(31)		
(32)		
(33)	←	Record mode, drum flip flop to internal FM modulator
(34)	→	To pin 12

(Continued overleaf)

PJ22031A

(35) 250mV pk-pk from pin 7
(36)
(37) 1v0
(38) 0v
(39) 4v0
(40) ← 5v Supply
(41) 700mV pk-pk luminance waveform
(42)
(43) 2v9
(44) → Record 1v0 pk-pk FM to video head circuit

PU22046A

Chroma playback and record processor — Package type B (32 pins)

Pins		Function
(1) 2v5		
(2) 2v5	←	0v2 pk-pk, 627kHz Playback chroma from head amplifier
(3) 0v Ground		
(4) 2v3 playback, 5v0 record		
(5)	→	800 mV pk-pk, 627kHz chroma to head amplifer
(6) 0v	←	5.06 MHz from pin 9 via bandpass filter
(7) 1v0	←	Detected chroma from pin 17
(8) 0v Ground		
(9)	→	5.06MHz to pin 6 via bandpass filter
(10) 400mV pk-pk 4.43MHz		
(11)	←	Drum flip flop pulse
(12) 5v0		
(13)	→	4.43MHz to servo circuits
(14) 4v0, 4.43Mhx xtal oscillator		
(15) 2v8, 4.43MHz, 0v5 pk-pk, xtal oscillator		
(16) 2v9		
(17) 2v4	→	Chroma detector to pin 7
(18) 2v8		
(19) 0v5	←	Video sync' pulse
(20) 2v8		
(21)	←	Mode control, Record = H, Playback = L
(22) 2v8	→	400mV pk-pk Playback chroma
(23) 2v7		
(24) 1v0	←	Playback chroma via delay line from pin 26
(25)	←	5v Supply
(26) 2v5	→	Playback chroma to pin 24 via delay line
(27) 0v Ground		
(28)	←	Secam transmission detector = H
(29)	←	Chroma,150mV pk-pk playback, 390mV pk-pk record from pin 31 via bandpass filter
(30) 2v5		
(31)	←	Chroma from pin 29 via bandpass filter
(32)	←	1v pk-pk E-E (Record) video

PU22199A

Luminance processor

Package type B
(44 pins)

Pins		Function
(1) 2v7	←	E-E level control
(2) 0v5	→	Composite sync'
(3) 2v8	←	1v0 pk-pk Video
(4) 2v2		
(5) 3v2	←	0v2 pk-pk playback luminance
(6) 1v5		
(7) 1v6	→	Luminance to low pass filter
(8) 0v Ground		
(9) 2v2	→	E-E or playback video
(10) 2v8	←	0v4 pk-pk Playback chroma
(11) 2v7		
(12) 3v3		
(13) 1v2		
(14) 0v	←	Vertical pulse
(15) 1v0		
(16) 1v4	→	0v5 pk-pk Demodulated playback FM
(17)		
(18) 0v Ground		
(19)	←	5v Supply
(20) 5v0	←	Luminance delay line
(21) 4v0	→	Luminance delay line
(22) 3v2	←	0v4 pk-pk playback FM
(23)		
(24) 3v0	→	0v3 pk-pk Luminance to pin 5
(25) 2v6	←	Picture sharpness control
(26)		
(27) 1v2	←	Drum flip flop Pulse
(28) 3v5		
(29)	←	Luminance from pin 16
(30) 3v5		
(31) 3v0		
(32) 1v0		
(33) 2v3	←	Drum flip flop pulse
(34)		
(35) 1v5	←	0v3 pk-pk Luminance from pin 7 via low pass filter
(36) 2v3	→	Luminance to pin 39
(37) 0v9		

(Continued opposite)

(38) 0v SP
(39) 4v1
(40) ← 5v Supply
(41)
(42)
(43) 2v9, White clip adjustment
(44) 4v0 → 1v pk-pk Record FM

PU22282

Luminance processor

Package type B (44 pins)

Pins		Function
(1) 2v5		
(2) 0v4	→	Composite sync' pulses
(3) 2v8	←	PB & E-E Video
(4) 2v8		
(5) 3v3	←	PB video from pin 24
(6) 1v2		
(7) 1v6	→	560mV pk-pk record mode video to pin 35 via low pass filter
(8) 0v Ground		
(9) 1v9	→	Composite video
(10) 2v4	←	0v4 pk-pk Playback Chroma
(11) 2v4		
(12) 3v0		
(13) 1v5		
(14) 0v	←	Vertical pulse
(15) 0v4		
(16) 1v5	→	0v4 pk-pk Playback Demodulated video
(17)		
(18) 0v Ground		
(19)	←	5v Supply, Playback mode
(20) 5v0		
(21) 5v0		
(22) 0v4	←	FM from Head pre-amplifier
(23) 5v0		
(24) 2v9	→	Video from internal sharpness control circuit to pin 5
(25) 2v1	←	Sharpness control
(26) 3v8		
(27) 1v4		
(28) 3v7		
(29) 3v8	←	Demodulated video from pin 16
(30) 3v9		
(31) 3v1		
(32) 1v0		
(33) 2v2		
(34) 0v2		
(35)	←	Record mode video from pin 7
(36)	→	Record mode video to detail enhancer circuit

(Continued opposite)

(37) 0v7
(38) ← SP mode = Lo
(39) 3v9 ← Record mode video from detail enhancer circuit
(40) ← 5v Supply
(41) 2v2
(42) 2v1
(43) 2v0
(44) 3v3 → Record FM to head pre-amplifier

PU321A

Capstan motor drive output

Package type A (8 pins)

Pins			Function
(1)		←	18v Supply
(2)	15v9	←	Drive control 1
(3)	8v8	→	Capstan motor drive 1
(4)	15v9	←	Drive control 2
(5)	8v8	→	Capstan motor drive 2
(6)	15v8	←	Drive control 3
(7)	8v8	→	Capstan motor drive 3
(8)		←	18v Supply

PU4310

Reel motor drive

Package type A (10 pins)

Pins		Function
(1)	←	–ve supply
(2)	←	Motor rotation control
(3)	→	Motor –ve supply
(4)	←	Motor rotation control
(5)	→	Motor +ve supply
(6)	←	Motor rotation control
(7)	→	Motor +ve supply
(8)	←	Motor rotation control
(9)	←	Motor –ve supply
(10)	←	+ve supply

Internally the device is a two PNP & two NPN transistor array.

SA2006A

Capstan motor drive — **Package type B (30 pins)**

Pins		Function
(1)	←	+ve Hall effect device 1
(2)	←	–ve Hall effect device 1
(3)	←	+ve Hall effect device 2
(4)	←	–ve Hall effect device 2
(5)		
(6) Ground		
(7) Ground		
(8) Ground		
(9) Ground		
(10)	←	14v Supply
(11)	→	Stator coil 1 drive
(12)	→	Stator coil 2 drive
(13)	→	Stator coil 3 drive
(14)	→	Output current sensing & feedback to pin 26
(15) n/c		
(16)	←	5v Supply
(17)	←	FG pulses
(18)	←	FG pulses
(19)	→	FG pulse output
(20) Ground		
(21)	←	Reference input voltage from servo drive
(22) Ground		
(23) Ground		
(24) Ground		
(25)	←	Drive from servo
(26)	←	From pin 14
(27)	←	Forward / reverse control
(28) n/c		
(29)	←	+ve Hall effect device 3
(30)	←	–ve Hall effect device 3

SD3624A

Drum & Capstan servo

Package type B (42 pins)

Pins		Function
(1)	⟶	4v0 pk-pk Record mode ACE control pulses
(2) 0v		
(3)		
(4)	⟵	1v5 pk-pk ramp, playback tracking control adjustment
(5) 5v0		
(6) 5v0		
(7) 5v1		
(8) 0v7	⟵	4v0 pk-pk, Record mode composite sync' pulses
(9)	⟵	4v0 pk-pk 25Hz Head switching flip flop pulse
(10)		
(11)	⟶	Vertical sync' pulse output
(12)	⟵	5v Supply
(13) 0v		
(14) 0v1		
(15) 0v	⟵	Head 2 switching adjustment
(16) 0v	⟵	Head 1 switching adjustment
(17) 2v4	⟵	Drum PG pulse
(18)	⟵	Drum FG pulse
(19)	⟵	4v0 pk-pk, Playback Capstan ACE head control pulse
(20)	⟵	4v0 pk-pk squarewave, Capstan FG pulse
(21) 0v Ground		
(22) 0v7	⟵	150mV pk-pk, 4.43MHz clock input
(23) 2v5		
(24)	⟶	Squared ACE head control pulse output
(25)	⟶	Capstan FG pulse output
(26) 5v0	⟵	Capstan speed select control
(27) 5v0	⟵	Capstan speed select control
(28)	⟵	Vertical pulse on / off control
(29)		
(30)		
(31)	⟵	Capstan forward = L, Capstan reverse = H
(32) 0v		
(33)		
(34)	⟶	4v0 pk-pk squarewave, Drum motor drive output
(35)	⟶	4v0 pk-pk squarewave, Drum phase control output
(36) 2v5	⟶	5v0 pk-pk squarewave, Capstan phase control to pin 40

(Continued overleaf)

SD3624A

(37) ⟶ 5v0 pk-pk squarewave, Capstan speed control to pin 39
(38) ⟵ 5v Supply
(39) ⟵ From pin 37
(40) ⟵ From pin 36
(41) ⟶ Capstan motor drive output
(42) 0v Ground

SL486

Infra red receiver

Package type B
(16 pins)

Pins		Function
(1)	←	Cathode of infra red diode
(2)		
(3)		
(4)		
(5)		
(6)		
(7)	←	5v Supply
(8)		
(9)		
(10)		
(11)	→	Received data to micro'
(12) 0v Ground		
(13) 0v Ground		
(14) 0v Ground		
(15)		
(16)	→	Anode of infra red diode

STA321A

Capstan motor drive

Package type A (8 pins)

Pins			Function
(1)		←	18v Supply
(2)	15v9	←	Drive control 1
(3)	8v8	→	Capstan motor stator drive 1
(4)	15v9	←	Drive control 2
(5)	8v8	→	Capstan motor stator drive 2
(6)	15v8	←	Drive control 3
(7)	8v8	→	Capstan motor stator drive 3
(8)		←	18v Supply

STA401A

4 × Invertors

Package type A
(10 pins)

Pins		Function
(1) n/c		
(2)	←	Invertor 1 input
(3)	→	Invertor 1 output
(4)	←	Invertor 2 input
(5)	→	Invertor 2 output
(6)	←	Invertor 3 input
(7)	→	Invertor 3 output
(8)	←	Invertor 4 input
(9)	→	Invertor 4 output
(10) 0v Ground		

Invertors designated numbers 1,2, etc are only given to distinguish between the individual devices

STK5322

Power supply regulator — **Package type A (10 pins)**

Pins		Function
(1) 0v Ground		
(2)	⟶	Regulated 9v5
(3) 12v5		
(4)	⟵	15v Supply
(5)	⟶	Regulated 15v0
(6) 18v1 Bias		
(7)	⟵	18v Supply
(8) n/c		
(9)	⟵	Power control, On = L (0v5)
(10) 0v Ground		

STK5326

Power supply regulator

Package type A (10 pins)

Pins		Function
(1) 0v Ground		
(2)	→	12v regulated output
(3)		
(4)	←	18v Supply
(5)	→	13v regulated output
(6)	←	18v Supply
(7)	←	18v Supply
(8)	←	Pin 2 output, on / off control
(9)	←	Pin 5 output, on / off control
(10) 0v Ground		

STK5331

Power supply regulator

Package type A
(8 pins)

Pins		Function
(1)	⟶	Switched 5v output
(2)	⟵	6v unregulated input
(3)	Connected to a load resistor	
(4)	Connected to a load resistor	
(5)	⟶	Switched 12v output
(6)	⟵	On / off control input
(7)	⟵	Up to 30v Unregulated input
(8)	0v Ground	

STK5332

Power supply regulator **Package type A (8 pins)**

Pins		Function
(1)	⟶	Switched 5v output
(2)	⟵	6v input via external switching transistor
(3)	Connected to a load resistor & ⟶	on / off drive to external switching transistor
(4)	Connected to a load resistor	
(5)	⟶	Switched 13v output
(6)	⟵	On / off control input
(7)	⟵	Up to 30v unregulated input
(8)	0v Ground	

STK5333

Power supply regulator **Package type A (8 pins)**

Pins		Function
(1) 0v Ground		
(2)	⟶	5v regulated output
(3)	⟶	Always 6v output
(4)		
(5)	⟵	DC input
(6)	⟶	15v regulated output
(7)	⟵	Power on / off control input, on = Lo
(8)	⟵	Unregulated DC supply

STK5338

Power supply regulator

Package type A
(8 pins)

Pins		Function
(1)	0v Ground	
(2)	⟶	5v regulated output
(3)	⟶	5v regulated output
(4)	⟵	30v input
(5)	Connected to a load resistor	
(6)	⟶	12v regulated output
(7)	⟵	Regulated outputs on = H
(8)	⟵	30v input

STK5339

Power supply regulator

Package type A (8 pins)

Pins		**Function**
(1)	0v Ground	
(2)	⟶	Regulated 5v output
(3)	⟶	Regulated 6v output
(4)	⟵	Up to 30v Unregulated input
(5)	Connected to a load resistor	
(6)	⟶	Regulated 12v output
(7)	⟵	On / off control input
(8)	⟵	Up to 30v Unregulated input

STK5340

Power supply regulator

Package type A (8 pins)

Pins		Function
(1)	←	Unregulated 14v input
(2)	←	On / pff control, On = H
(3)	→	Regulated 12v output
(4) Connected to load resistor		
(5)	←	Unregulated 8v5 input
(6)	→	Regulated 6v output
(7)	→	Regulated 5v output
(8) 0v Ground		

STK5352

Power supply regulator

Package type A (10 pins)

Pins		**Function**
(1)	0v Ground	
(2)	⟶	Regulated 6v0
(3)	n/c	
(4)	⟵	12v Supply
(5)	⟶	Regulated 12v0
(6)	14v9 Bias	
(7)	⟵	16v Supply
(8)	0v Ground	
(9)		
(10)		

STK5372

Power supply regulator

Package type A (8 pins)

Pins		Function
(1)	⟶	Regulated 5v output
(2)	⟵	Shut down
(3)	⟵	Up to 30v unregulated input
(4)	⟶	Regulated 12v output
(5)	⟶	Regulated12v output
(6)	Connected to a load resistor	
(7)	⟵	Up to 30v unregulated input
(8)	0v Ground	

STK5391

Power supply voltage regulator

Package type A (9 pins)

Pins		Function
(1)	0v Ground	
(2)	→	Regulated 5v
(3)	→	Always 5v
(4)	←	8v Supply
(5)	←	24v Supply
(6)	→	Regulated switched 12v
(7)	←	Power on = H
(8)	←	14v Supply
(9)	←	44v Supply

STK5392

Power supply regulator

Package type A
(9 pins)

Pins		Function
(1)	←	45v unregulated
(2)	←	14v
(3)	←	Power on / off control
(4)	→	12v regulated
(5)		
(6)	←	7v
(7)	→	5v regulated
(8)	→	5v regulated
(9)	0v Ground	

STK5421

Power supply regulator

Package type A (15 pins)

Pins		Function
(1)	⟶	Regulated 12v output
(2)	Connected to a load resistor	
(3)	⟶	Regulated12v output
(4)		
(5)	⟵	Up to 30v unregulated input
(6)	Connected to pin 5	
(7)	⟵	Up to 30v unregulated input
(8)	⟶	Regulated16v output
(9)	Connected to pin 8	
(10)	⟶	Regulated 9v output
(11)	Connected to a load resistor	
(12)	0v Ground	
(13)	⟵	On / off control input
(14)		
(15)		

STK5422

Power supply regulator

Package type A (15 pins)

Pins		**Function**
(1)	⟶	Regulated 12v output
(2)	Connected to a load resistor	
(3)	⟶	Regulated 12v output
(4)		
(5)	⟵	Up to 30v unregulated input
(6)	Connected to pin 5	
(7)	⟵	Up to 30v Unregulated input
(8)	⟶	Regulated 13v output
(9)	Connected to pin 8	
(10)	⟶	Regulated 9v output
(11)	Connected to a load resistor	
(12)	0v Ground	
(13)	⟵	On / off control input
(14)		
(15)		

STK5431

Power supply regulator

Package type A (15 pins)

Pins		Function
(1)	⟶	Regulated 5v output
(2)		
(3)	⟶	Regulated 12v output
(4)		
(5)	⟵	Up to 30v unregulated input
(6)	Connected to pin 3	
(7)	⟵	Up to 30v unregulated input
(8)	⟶	Regulated 15v output
(9)	Connected to pin 8	
(10)	⟶	Regulated 9v output
(11)	Connected to a load resistor	
(12)	0v Ground	
(13)	⟵	On / off control input
(14)		
(15)		

STK5434

Regulated power supply

Package type A (15 pins)

Pins		Function
(1)	→	5v Regulated output
(2)		
(3)	→	12v Regulated output
(4) 16v		
(5)	←	23v Unregulated
(6)	←	12v Unregulated
(7)	←	22v Unregulated
(8)	←	22v Unregulated
(9)	→	16v Regulated output
(10)	→	9v Regulated output
(11) 10v		
(12) 0v Ground		
(13)	←	Pin 9 & pin 10 on = H
(14) 0v6		
(15) 9v4		

Unregulated voltages will vary

STK5451

Power supply regulator

Package type A (15 pins)

Pins		Function
(1)	——→	Switched 12v output
(2)	Connected to a load resistor	
(3)	←——	Up to 30v unregulated input
(4)	←——	Connected to a load resistor
(5)	——→	Switched 5v output
(6)	Connected to Bias circuitry	
(7)	——→	12v or 16v Switched output
(8)	Connected to a load resistor	
(9)	Connected to pin 3	
(10)	←——	Up to 30v Unregulated input
(11)	——→	M start
(12)	0v Ground	
(13)	←——	On/off control input
(14)	——→	On /off drive to external switching transistor
(15)	Connected to a load resistor	

STK5461

Power supply regulator

Package type A (15 pins)

Pins		Function
(1)	⟶	Regulated 5v output
(2)		
(3)		
(4)		
(5)		
(6)	⟵	Up to 30v Unregulated input
(7)	⟵	Up to 30v Unregulated input
(8)	Connected to pin 7	
(9)	⟶	Regulated 12v output
(10)	⟶	Regulated 12v output
(11)	Connected to a load resistor	
(12)	0v Ground	
(13)	⟵	Switch off Pin 9 output
(14)	⟵	Switch off Pin 10 output
(15)	Connected to a load resistor	

STK5466

Power supply regulator **Package type A (15 pins)**

Pins		Function
(1)	⟶	Regulated 5v output
(2)		
(3)		
(4)		
(5)		
(6)	⟵	Up to 30v unregulated input
(7)	⟵	Up to 30v unregulated input
(8) Connected to pin 7		
(9)	⟶	Regulated 12v output
(10)	⟶	Regulated 12v output
(11)		
(12) 0v Ground		
(13)	⟵	Pins 9 & 10 outputs cut off
(14) Connected to pin 13		
(15) Connected to a load resistor		

STK5467

Power supply regulator

Package type A (12 pins)

Pins		Function
(1)	⟶	Regulated 5v output
(2)		
(3)		
(4)		
(5)	⟵	Up to 30v unregulated input
(6)	⟵	Up to 30v unregulated input
(7)	Connected to pin 6	
(8)	⟶	Regulated 12v output
(9)	⟶	Regulated 12v output
(10)	⟵	Pin 9 output cut off
(11)	0v Ground	
(12)	Connected to a load resistor	

STK5471

Power supply regulator

Package type A (10 pins)

Pins		Function
(1)	⟶	Regulated 5v output
(2)	⟵	Regulated outputs cut off
(3)	⟵	Up to 30v unregulated input
(4)	⟶	Regulated 12v output
(5)	Connected to a load resistor	
(6)	⟵	Up to 30v unregulated input
(7)	⟶	Regulated 12v output
(8)	Connected to a bias resistor	
(9)	Connected to pin 6	
(10)	0v Ground	

STK5473

Power supply regulator

Package type A (10 pins)

Pins		**Function**
(1)	0v Ground	
(2)	⟶	5v7
(3)	⟶	12v
(4)	2v9 Mains Transformer centre tap	
(5)	⟶	12v
(6)	⟵	Unregulated 13v5
(7)	⟵	Unregulated16v
(8)	⟶	13v
(9)	0v	
(10)	⟵	Unregulated14v

STK5481

Power supply regulator

Package type A (12 pins)

Pins		Function
(1)	←	On / off control to pin 3 output
(2)	←	Up to 30v unregulated input
(3)	→	Regulated 5v output
(4)	→	Regulated 12v output
(5)	←	On / off control to pin 4 output
(6)	→	Regulated 12v output
(7)	Connected to a load resistor	
(8)	←	Up to 30v unregulated input
(9)	→	Regulated 12v output
(10)	Connected to a load resistor	
(11)	Connected to pin 8	
(12)	0v Ground	

STK5482

Power supply regulator | **Package type A (15 pins)**

Pins		Function
(1)	0v Ground	
(2)	←	Up to 30v unregulated input
(3)	→	Regulated 5v output
(4)	→	Regulated 12v output
(5)	←	Bias circuit & On / off control to pin 4 output
(6)	←	Up to 30v unregulated input
(7)	→	Regulated 5v output
(8)	←	Bias circuit & On / off control to pin 7 output
(9)	Connected to pin 2	
(10)	Connected to a load resistor & pin 7 output on /off switching	
(11)	→	Regulated 13v & 15v output
(12)	←	Switching control to pin 11
(13)	Connected to pin 6	
(14)	←	Switching control to pin 11
(15)	0v Ground	

STK5490

Power supply regulator

Package type A (12 pins)

Pins		Function
(1) 17v		
(2)	←	Unregulated 9v supply
(3)	→	Regulated 5v
(4)	→	Regulated 12v
(5)	←	Power on control =L
(6)	→	Always 12v
(7)		
(8)	←	Unregulated 18v supply
(9)	→	12v motor supply
(10) 13v0		
(11)	←	Unregulated 18v
(12) 0v Ground		

STK6962

Capstan motor drive

Package type A
(7 pins)

Pins		**Function**
(1)	0v Ground	
(2)	←	Mode control
(3)	→	Capstan motor drive
(4)	←	12v Supply
(5)		
(6)	→	Capstan motor drive
(7)	←	Mode control

STK7348

Chopper type Power supply regulator

Package type A (10 pins)

Pins		Function
(1)	39v	
(2)	0v	
(3)	0v2	
(4)	32v	Control feedback
(5)	–ve 0v7	
(6)	0v	
(7)	0v2	← initial start up bias supply
(8)	0v3	→ Output Current sensing
(9)	0v3	
(10)		→ DC supply & Chopper output to transformer primary

STR10006C

Power supply chopper & control

Package type A (5 pins)

Pins		Function
(1)		
(2)	←	Start up & control source
(3)	→	Output to chopper transformer primary winding
(4)	→	Chopper output current sensing resistor
(5)	←	–ve supply

STR1096

Power supply regulator

Package type A (5 pins)

Pins		Function
(1)	←	15v Supply
(2)	←	On / off control, On =10v, Off = Lo
(3)	0v Ground	
(4)	→	Switched 9v
(5)	→	Switched 5v

STR9012

Power supply regulator

Package type A (5 pins)

Pins		Function
(1)	——→	12v Regulated
(2)	←——	Output voltage control
(3)	←——	Output on / off control
(4)	←——	(17v typical),unregulated supply from bridge rectifier
(5) 0v Ground		

STRD6008

Switched mode supply control & output

Package type A (5 pins)

Pins			Function
(1)	–ve 4v0	←	Control input
(2)	–ve 3v0	←	Start up & bias
(3)		→	Switched output to chopper Transformer primary
(4)		←	–ve bridge rectifier supply
(5)		←	Over voltage control

STRD6802

Power supply regulator

Package type A
(5 pins)

Pins		Function
(1)	0v ground	
(2)	←	Switch on bias
(3)	→	Output to chopper transformer primary
(4)	→	Current sensing circuit
(5)	←	Drive control

STV5722

Video head preamplifier **Package type B (16 pins)**

Pins		Function
(1)	⟶	Record FM to head (grounded during PB mode)
(2)		
(3)		
(4)		
(5)	⟵	Playback FM from head 2
(6) 0v Ground		
(7)	⟵	Playback FM from head 1
(8)		
(9)	⟵	Head switching flip flop pulse
(10)	⟵	5v Supply
(11)	⟶	Playback FM output
(12) 0v Ground		
(13)	⟵	Composite sync' pulse
(14)	⟶	Playback video envelope
(15)	⟵	Record mode Luminance & chroma
(16)		

STV5726

Video head preamplifier

Package type B (20 pins)

Pins		Function
(1)	⟶	Record mode FM to SP heads
(2)	Grounded during record mode, SP head 2	
(3)	Grounded during record mode, SP head 1	
(4)	⟵	SP, FM Playback from head 2
(5)	⟵	SP, FM Playback from head 1
(6)	0v Ground	
(7)	⟵	LP, FM Playback from head 1
(8)	⟵	LP, FM Playback from head 2
(9)	Grounded during record mode, LP head 1	
(10)	Grounded during record mode, LP head 2	
(11)	⟶	Record mode FM to LP heads
(12)	⟶	LP / SP playback detection
(13)	⟵	Head switching flip flop pulse
(14)	⟵	5v Supply
(15)	⟶	Playback FM output
(16)	0v Ground	
(17)	⟵	Composite sync'
(18)	⟶	Playback envelope detection
(19)	⟵	Record mode luminance & chroma
(20)		

TA4168

Infra red remote receiver & separate digital to analogue convertor

Package type A (14 pins)

Pins		Function
(1) 0v Ground		
(2)	←	Data
(3) 0v Ground		
(4)	←	Data
(5)	←	Data
(6) 0v Ground		
(7)	←	Data
(8)	→	Analogue output
(9)	←	9v Supply
(10)	→	Remote control data out
(11) 3v4		
(12) 3v0		
(13)	←	Infra red receiver diode input
(14) 0v Ground		

TA7245BP

Drum motor drive

Package type B
(14 pins)

Pins		**Function**
(1) 6v0	←	Hall effect device 1
(2) 6v0	←	Hall effect device 1
(3) 6v0	←	Hall effect device 2
(4) 6v0	←	Hall effect decice 2
(5) Playback mode = 0v4, Stop mode = 0v		
(6) 10v	→	Stator coil drive 1
(7) 10v	→	Stator coil drive 2
(8) 10v	→	Stator coil drive 3
(9)	←	14v Supply
(10) 0v5		
(11) 4v0	←	Control
(12) 4v5	←	Control
(13) 6v0	←	Hall effect device 3
(14) 6v0	←	Hall effect device 3

TA7267

Reel motor drive

Package type A (7 pins)

Pins		Function
(1)	←	Motor direction control logic, Forward = L, Rewind = H
(2)	←	Motor direction control logic, Forward = H, Rewind = L
(3)	→	Motor –ve terminal drive
(4)	←	Motor current limit control
(5)	→	Motor +ve terminal drive
(6)	←	Full speed fast forward / rewind control
(7)	←	5v Supply

TA7288P

Motor control driver

Package type A (10 pins)

Pins		**Function**
(1)	Common motors 0v	Ground
(2)	⟶	Loading motor –ve terminal
(3)	⟶	Carriage motor –ve terminal
(4)	⟵	Mode control
(5)	⟵	Mode control
(6)	⟵	Mode control
(7)	⟵	5v Supply
(8)	⟵	Loading control
(9)	⟵	Motors 5v Supply
(10)	⟶	Loading & Carriage motors common +ve terminal

Above example shows the device controlling two motors via the control inputs on pins 4,5 & 6

Control logic
Stop mode: Pin 4,5 & 6 = H
Loading mode: Pin 4 = H, pins 5 & 6 = L
Unloading mode: Pin 4= H, Pin 5 = L, Pin 6 = H

TA7291S

Loading motor drive

Package type A (9 pins)

Pins		Function
(1)	→	Forward mode control
(2)	←	5v Supply
(3)	→	14v Forward drive to motor
(4) n/c		
(5) 0v Ground		
(6)	←	14v Supply
(7)	→	14v Reverse drive to motor
(8)	←	14v Supply
(9)	←	Reverse mode control

TA7325

Dual operational audio amplifiers

Package type A (9 pins)

Pins		Function
(1)	←	Amplifier 1 non inverted input
(2)	←	Amplifier 1 inverted input
(3)	→	Amplifier 1 output
(4)	←	12v Supply
(5) 0v Ground		
(6)	→	Amplifier 2 output
(7)	←	Amplifier 2 inverted input
(8)	←	Amplifier 2 non inverted input
(9) 0v Ground		

Amplifiers designated numbers, i.e 1,2 are given to distinguish between the individual devices

TA7337P

Sound I.F detector

Package type A (9 pins)

Pins		Function
(1) n/c		
(2) 1v7	←	40mV pk-pk Sound I.F input
(3) 1v7		
(4)	←	12v Supply
(5) 0v Ground		
(6) 4v7, FM detector LC circuit		
(7) 4v7, FM detector LC circuit		
(8) 4v1		
(9) 3v5	→	1v0 pk-pk Audio output

TA7348P

3 input to 1 output switch

Package type A (9 pins)

Pins		Function
(1)	←	Input 1
(2)	←	Logic control
(3)	←	Input 2
(4)	←	Logic control
(5) 0v Ground		
(6)	←	Input 3
(7)	←	Mute input
(8)	→	Output
(9)	←	9v Supply

TA7355P

2 × single pole switches sharing a common pole **Package type A (7 pins)**

Pins		**Function**
(1)	←	Switch 1 input
(2)	→	Switches 1 & 2 common pole output
(3)	←	Switch 2 input
(4)	0v Ground	
(5)	←	To close switch 2, (pin 2 to pin 3) = H
(6)	←	12v Supply
(7)	←	To close switch 1, (pin 1 to pin 2) = H

For clarity the signal output is from the common pole. In practice, the signal may be in any direction

TA7361

Dual one way single pole switch

Package type A (7 pins)

Pins	Function
(1)	Switch 1 pole
(2)	Switch 1 & 2 common connection
(3)	Switch 2 pole
(4)	0v Ground
(5)	← Switch 1 control. Make = H, Open = L
(6)	← 12v Supply
(7)	← Switch 2 control. Make = H, Open = L

Although both switches have a common connection inside the device. It is possible, due to the independent switching facility, to route a signal into pin1 or pin 3 and output both at pin 2 (added together), or individually.

TA75339

4 × Operational amplifiers

Package type B (14 pins)

Pins		Function
(1)	→	Amplifier 1 output
(2)	→	Amplifier 2 output
(3)	←	12v Supply
(4)	←	Amplifier 2, inverted input
(5)	←	Amplifier 2, non inverted input
(6)	←	Amplifier 1, inverted input
(7)	←	Amplifier 1, non inverted input
(8)	←	Amplifier 3, inverted input
(9)	←	Amplifier 3, non inverted input
(10)	←	Amplifier 4, inverted input
(11)	←	Amplifier 4, non inverted input
(12) 0v Ground		
(13)	→	Amplifier 4 output
(14)	→	Amplifier 3 output

Amplifier designated numbers, i.e 1,2, etc are only given to distinguish between the individual devices

TA75358P

Dual operational amplifiers **Package type B (8 pins)**

Pins		Function
(1)	→	Amplifier 1 output
(2)	←	Amplifier 1 Inverted input
(3)	←	Amplifier 1 Non inverted input
(4) 0v Ground		
(5)	←	Amplifier 2 non inverted input
(6)	←	Amplifier 2 inverted input
(7)	→	Amplifier 2 output
(8)	←	31v Supply voltage

1 × amplifier often used in RF circuits as a tuning voltage buffer. Hence the high supply voltage.
The other amplifier (normally amplifier 2) not connected.

TA75802P

4 × operational amplifiers **Package type B (14 pins)**

Pins		Function
(1)	→	Amplifier 1 output
(2)	←	Amplifier 2, inverted input
(3)	←	Amplifier 1, non inverted input
(4)	←	12v Supply
(5)	←	Amplifier 2, non inverted input
(6)	←	Amplifier 2, inverted input
(7)	→	Amplifier 2 output
(8)	→	Amplifier 3 output
(9)	←	Amplifier 3 inverted input
(10)	←	Amplifier 3, non inverted input
(11) 0v Ground		
(12)	←	Amplifier 4, non inverted input
(13)	←	Amplifier 4, inverted input
(14)	→	Amplifier 4 output

Amplifiers designated numbers, i.e 1,2, etc are given to distinguish between the individual devices

TA75902

4 × Operational amplifiers

Package type B (14 pins)

Pins		Function
(1)	→	Amplifier 1 output
(2)	←	Amplifier 1, inverted input
(3)	←	Amplifier 1, non inverted input
(4)	←	12v Supply
(5)	←	Amplifier 2, non inverted input
(6)	←	Amplifier 2, inverted input
(7)	→	Amplifier 2 output
(8)	←	Amplifier 3 output
(9)	←	Amplifier 3, inverted input
(10)	←	Amplifier 3, non inverted input
(11)	0v Ground	
(12)	←	Amplifier 4, non inverted input
(13)	←	Amplifier 4, inverted input
(14)	→	Amplifier 4 output

Amplifier designated numbers, i.e 1,2, etc are given to distinguish between the individual devices

TA7607AP

Vision I.F demodulator — **Package type B (16 pins)**

Pins			Function
(1)	5v0	←	I.F from SAW Filter
(2)	5v0		
(3)	7v6	←	Tuner AGC manual adjustment
(4)	4v5	→	Automatic gain control (AGC) to tuner
(5)	6v6		
(6)	7v0 AFT, LC circuit		
(7)	4v0 AFC detector LC circuit		
(8)	8v0	→	Demodulated video & sound I.F to pin 9
(9)	8v0	←	From pin 8
(10)	4v4, AFC, LC circuit		
(11)		←	12v Supply
(12)	3v5	→	3v0 pk-pk, Demodulated video & sound I.F output
(13)	0v Ground		
(14)	8v0		
(15)	5v0		
(16)	5v0	←	I.F from SAW Filter

TA7629

Audio record & playback

Package type B (16 pins)

Pins			Function
(1)		←	9v Supply
(2)	4v5	←	Record & playback signal from pin 8
(3)	0v Ground		
(4)	4v4	←	Playback signal from audio head
		Record	signal from tuner or line input
(5)	4v3	→	Amplified pin 4 signal to pin 6
(6)	4v5	←	From pin 5
(7)		←	Record = L, playback = H
(8)		→	Record or playback audio output
(9)	4v6	→	Record audio to head circuit
(10)			
(11)		←	Noise reduction control
(12)	4v5		
(13)	4v4		
(14)	4v8		
(15)	5v0		
(16)	7v0		

TA7743

FM luminance playback drop out compensation detection

Package type B (18 pins)

Pins		Function
(1) 1v8	←	Trick speed picture drop out control
(2) n/c		
(3) 3v0	→	Drop out detected signal to pin 14
(4) 4v8		
(5) 0v Ground		
(6) 2v8	→	AGC controlled FM luminance to pin 17
(7) 3v8	←	FM luminance from video head circuit
(8) 2v8		
(9) 0v7		
(10) n/c		
(11) n/c		
(12)	←	5v Supply
(13) 1v6	→	Normal or drop out compensated FM output
(14) 2v9	←	From pin 3
(15) 3v2	→	Luminance from pin 13 via delay line (drop out compensation)
(16) 0v Ground		
(17) 3v2		
(18) n/c		

TA7772P

Video head preamplifier | **Package type B (16 pins)**

Pins		Function
(1)	0v Ground	
(2)	←	Playback FM from head 1
(3)	→	2v7 pk-pk FM record mode to heads
(4)	0v Ground	
(5)	→	2v7 pk-pk FM record mode to heads
(6)	←	Playback FM from head 2
(7)	←	5v playback mode supply
(8)	0v Ground	
(9)		
(10)	→	300mV pk-pk playback FM
(11)	←	350mV pk-pk record mode FM
(12)	←	9v record mode supply
(13)	0v Ground	
(14)	←	25 Hz Head switching flip flop pulse
(15)	←	350mV pk-pk record mode FM
(16)		

Pins 3 & 5 may be linked together

TA8400P

Loading motor & carriage motor control

Package type B (16 pins)

Pins		Function
(1)		
(2)	←	Mode motor forward = H, Reverse = H Carriage motor forward = L, Reverse = L Brake = H, Stop = L
(3)	←	Mode motor forward = L, Reverse = L Carriage motor forward = H, Reverse = H Brake = H, Stop = L
(4) 0v Ground		
(5) 0v Ground		
(6)	←	Mode motor forward = L, Reverse = H Carriage motor forward = L, Reverse = H Brake & Stop = Intermediate level
(7)	←	12v Supply
(8)		
(9)	←	12v Supply
(10)	→	Mode motor forward = open, Reverse = open Carriage motor forward = L, Reverse = H Brake = L, Stop = open
(11)	→	Mode motor forward = H, Reverse = L Carriage motor forward = H, Reverse L Brake = L, Stop = open
(12) 0v Ground		
(13) 0v Ground		
(14)		
(15)	→	Mode motor forward = L, Reverse = H Carriage motor forward = open, Reverse = open Brake = L, Stop = open
(16)	←	12v Supply

TA8405S

Carriage & mode motor drive

Package type A (9 pins)

Pins		Function
(1)	←	Control from System i.c
(2)	←	Control from system i.c
(3)	←	Control from system i.c
(4)	→	Drive to Carriage motor, Forward = H, Reverse = L Mode motor, Forward = H, Reverse = L Brake = L, Stop = Open
(5)	0v Ground	
(6)	→	Drive to Carriage motor, Forward = L, Reverse = H Mode motor, Forward = L, Reverse = H
(7)	→	Drive to Carriage motor, Forward = L, Reverse = H Mode motor, Forward = Open, Reverse = Open Brake = L, Stop = H
(8)	←	12v Supply
(9)	←	12v Supply

TA8605N

Luminance FM modulator & demodulator

Package type B (30 pins)

Pins		Function
(1) 2v0	⟶	Demodulated playback video to external DOC circuit or record mode video from pin 4 via internal agc / clamp circuits
(2)		
(3) 2v3		
(4)	⟵	1v pk-pk record mode video
(5)		
(6)	⟵	520mV pk-pk playback video from external DOC & sharpness circuits
(7)	⟵	Vertical sync' pulse
(8)	⟵	Chroma
(9) 0v Ground		
(10)	⟶	Luminance & chroma to RF output modulator
(11)		
(12) 2v2	⟵	Record mode video from pin 1 via external limiter & comb filter circuits
(13)	⟵	Playback mode = 4v4, record mode = 3v0
(14)	⟵	Dark clip control
(15)		
(16)	⟶	1v5 pk-pk record mode video from intenal main emphasis circuit to pin 18
(17)	⟵	400mV pk-pk playback FM
(18)	⟵	Record mode video from pin 16 via deviation control
(19)	⟵	ac coupled to pin 17
(20)	⟵	5v Supply
(21)	⟶	Record mode modulated FM to head circuit
(22) 0v Ground		
(23)		
(24)		
(25)	⟶	4v4 pk-pk vertical sync' pulses to chroma circuit
(26) 1v5		
(27)		
(28) 0v8		
(29) 2v8		
(30)		

Playback FM amplifier, AGC & drop out detector

Package type B (16 pins)

Pins			Function
(1)	0v8		
(2)	2v8	←	FM video from pin 3
(3)		→	300mV pk-pk FM to pin 2
(4)			
(5)			
(6)	0v Ground		
(7)	3v2		
(8)			
(9)	3v2	←	300mV pk-pk FM from head preamplifier stage
(10)		←	5v Supply
(11)		→	1v pk-pk FM to pin 12
(12)		←	FM from pin 11
(13)	0v Ground		
(14)	2v3		
(15)		→	Signal drop out detection = L
(16)	2v0	→	1v2 pk-pk FM to demodulator circuit

TA8611AN

Sound & vision I.F detector

Package type B (20 pins)

Pins		Function
(1) 5v8		
(2) 5v8		
(3) 3v0	←	Tuner AGC control
(4) 4v0	←	I.F from SAW Filter
(5) 4v0	←	I.F from SAW Filter
(6) 0v Ground		
(7) 4v0	→	AGC control to tuner
(8) 4v0	→	Sound I.F output
(9) 6v0, Sound FM detector LC circuit		
(10) 6v0, Sound FM detector LC circuit		
(11)	←	12v Supply
(12) 2v5		
(13) 2v5	←	Sound I.F from pin 15 via 5.5MHz filter
(14) 0v Ground		
(15) 3v8	→	Detected video output
(16) 6v2, Vision demodulator LC circuit		
(17) 6v2, Vision demodulator LC circuit		
(18) 4v0, AFC, LC circuit		
(19)	←	12v Supply
(20) 5v0	→	AFC control to tuner

Pins 11 & 19 may be 9v supply

TA8617

ACE head control pulse & Capstan FG amplifier

Package type A (12 pins)

Pins		Function
(1) 2v5		
(2)	→	Capstan FG pulse to servo
(3)	←	5v Supply
(4)	→	Capstan ACE head phase control to servo
(5) 2v0	→	Feedback to pin 8
(6) 0v Ground		
(7) 2v0		
(8) 2v0	←	From pin 5
(9) 2v0	←	ACE head control pulse input
(10) 2v5		
(11) 2v6	←	Capstan FG pulse input
(12) 2v6		

TA8632N

Chroma processor | **Package type B (30 pins)**

Pins		Function
(1) 0V4	←	4v pk-pk Vertical sync' pulses
(2) 3v7	→	2v5 pk-pk vertical sync' pulse
(3) 5v, 5.06MHz voltage controlled oscillator RC circuit		
(4) 2v5		
(5) 2v6	←	4v pk-pk drum flip flop
(6) 4v6	→	7.16MHz to external i.c's
(7) 2v3		
(8) 2v5	→	4.43MHz xtal clock pulse to servo
(9)	←	5v Supply
(10) 3v0, 4.43MHz xtal		
(11) 0v Ground		
(12) 4v4,	→	400mV pk-pk sub convertor to pin 14 via 5.06MHz band pass filter
(13) 4v5		
(14) 2v6	←	From pin 12
(15)	←	Colour killer control
(16) 2v2	→	1v pk-pk record mode chroma
(17)	←	Speed select, SP = H, LP = L
(18) 2v4	←	1v pk-pk record & E-E video
(19) 2v9		
(20) 2v5	→	Chroma to pin 22 via 4.43MHz band pass filter
(21)	←	5v Supply
(22) 1v8	←	200mV pk-pk record mode chroma
(23) 0v Ground		
(24) 2v6	←	100mV pk-pk playback chroma
(25) 3v5		
(26)	←	800mV pk-pk playback chroma
(27)	←	System select, NTSC = H, SECAM = Medium, PAL = L
(28)	←	200mV pk-pk chroma from external comb filter
(29) 2v3		
(30) 2v8	→	800mV pk-pk playback chroma

TA8644

Chroma processor

Package type B (30 pins)

Pins		**Function**
(1)	←	Composite sync' pulses
(2)		
(3) 5v0		
(4) 1v8	←	Record & playback vertical lock sync' pulse from pin 29
(5) 4v4	←	25Hz flip flop pulse
(6) 5v0	→	8.86MHz pulses to drop out compensator delay line
(7) 2v0		
(8) 2v3, 4.43MHz xtal oscillator		
(9) 5v0		
(10) 3v0, 4.43MHz xtal oscillator		
(11) 0v Ground		
(12) 4v3	→	Record & playback sub convertor 5.06MHz to pin 14
(13) 3v0		
(14) 2v8	←	From pin 12
(15) 2v8		
(16)	→	Record mode 627kHz chroma to video head circuit
(17) 5v0		
(18) 2v5	←	Record mode video input
(19) 3v0		
(20) 2v5	→	Playback mode 4.43MHz chroma to pin 28 via comb filter
	→	Record mode 4.43MHz chroma to pin 22 via bandpass filter
(21)	←	5v Supply
(22) 2v7	←	From pin 20
(23) 0v		
(24)	←	Playback 627kHz chroma from video heads
(25)		
(26) 2v4		
(27)	←	PAL, Secam, NTSC select
(28) 2v7	←	From pin 20
(29)	→	To pin 4
(30)	→	Playback 4.43MHz chroma output

TA8757AN

Chroma processor **Package type B (24 pins)**

Pins			Function
(1)	2v4		
(2)	3v6, 4.43MHz xtal		
(3)	2v6		
(4)	3v7, 4.43MHz, xtal,800mV pk-pk		
(5)	3v0		
(6)	2v0		
(7)		←	5v Supply
(8)	2v4	→	400mV pk-pk record chroma
(9)	1v7		
(10)	3v0	←	800mV pk-pk chroma from pin 12
(11)	1v0		
(12)	2v4	→	4.43MHz chroma to pin 10 record mode. To luminance circuits playback mode
(13)	1v8	→	4.43MHz, 400mV pk-pk chroma to pin 17 playback mode.
(14)	3v3		
(15)	1v8	←	627kHz playback chroma from heads
(16)	5v2		
(17)	2v8	←	1v3 pk-pk composite video record mode. 400mV pk-pk 4.43MHz chroma from pin 13 playback mode
(18)	0v Ground		
(19)	1v5		
(20)	3v2	←	
(21)	0v3	←	Composite sync' record & playback modes.
(22)		←	5v Supply
(23)	1v2	←	1v5 pk-pk Drum flip flop pulse
(24)	2v5		

TA8871N

Luminance & chroma processor

Package type (?)
(54 pins)

Pins			Function
(1)		←	Tuner = H, External (line) = L
(2)	2v1		
(3)	0v Ground		
(4)		←	400mV pk-pk composite video
(5)	2v4	←	Picture control
(6)		→	2v8 pk-pk composite video
(7)			
(8)		→	600mV pk-pk Luminance to pin 10
(9)	2v4		
(10)		←	Luminance from pin 8
(11)		→	Playback mode 600mV pk-pk Luminance
		←	Record mode FM modulator
(12)		←	5v Supply
(13)		←	Playback mode 800mV pk-pk Luminance
		←	Record mode 900mV pk-pk Luminance
(14)	3v2	←	Playback level control
(15)		→	1v5 pk-pk playback Luminance to pin13 via Low pass filter
		→	1v3 pk-pk record mode composite video to pin 13
(16)			
(17)			
(18)		←	400mV pk-pk playback FM from heads
(19)			
(20)		→	200mV pk-pk, 8.8 MHz oscillator pulse
(21)			
(22)			
(23)	800mV pk-pk, 4.43 MHz		
(24)	2v2		
(25)	1v pk-pk, 4.43MHz xtal		
(26)			
(27)	2v5 pk-pk colour burst identification pulse		
(28)	1v4		
(29)		→	500mV pk-pk, 627kHz chroma to head circuit
(30)		←	5v Supply
(31)		←	PAL / SECAM control
(32)	2v6	←	External chroma
(33)	2v7		

(Continued overleaf)

TA8871N

(34) 2v1
(35) ← Record, Playback & Trick mode control
(36) → 2v0 pk-pk Burst gate pulse
(37) 1v5 ← 3v0 pk-pk Head switching pulse
(38) 3v0
(39)
(40) 2v5 ← 200mV pk-pk, 627kHz playback chroma
(41) 0v Ground
(42) 2v5 ← 4.43MHz, 200mV pk-pk chroma from pin 44 via comb filter
(43) ← LP = H, SP = L
(44) → 1v pk-pk chroma to pin 42
(45)
(46) 0v4 → 5v0 pk-pk Composite sync'
(47) 2v5
(48) 2v7 ← 800mV pk-pk Luminance from pin 50 via drop out compensator
(49) 3v0
(50) 2v3 → 800mV pk-pk Luminance to pin 48
(51) → 800mV pk-pk luminance, non linear playback de-emphasis & non linear record emphasis
(52) ← From pin 51
(53) 2v2
(54) 3v0

TA8892N

Luminance & chroma signal processor

Package type B (54 pins)

Pins		Function
(1) 2v9	←	On screen display video
(2)	→	2v5 pk-pk video to On screen display circuitry
(3) 1v6	←	Vertical & horizontal sync' blanking
(4)		
(5) 2v5		
(6) 2v7	→	700mV pk-pk to Playback equalization circuit
(7) 0v		
(8) 2v9	←	300mV pk-pk from playback equalization circuit
(9)	→	Record FM to video head preamplifier
(10) 0v Ground		
(11) 3v2	←	400mV pk-pk Playback FM from head circuit
(12)	←	5v Supply
(13)	←	IIC Data
(14)	←	IIC clock pulse
(15) 0v Ground		
(16)		
(17) 3v7		
(18) 2v0, 4.43MHz xtal		
(19) 2v8		
(20) n/c		
(21) 2v8		
(22) 4.43MHz xtal		
(23) 4v9		
(24)		
(25) 2v8		
(26)		
(27) 2v8	←	200mV pk-pk Playback chroma from delay circuit
(28)	←	5v Supply
(29)	→	300mV pk-pk Record chroma to video head preamplifier
(30)		
(31) 1v6	→	400mV pk-pk, playback chroma to pin 33 via NTSC system
(32) 1v8		
(33) 3v0	←	200mV pk-pk chroma from pin 31 (NTSC)
(34) 4v0		
(35) 2v6		

(Continued overleaf)

TA8892N

(36)
(37)
(38)
(39)
(40) ⟶ Burst gate switching pulse
(41) 0v4 ⟶ Horizontal sync' pulse
(42) ⟶ Composite sync' to head circuit
(43) 2v5
(44) 4v0 ⟵ Delayed Luminance from pin 47
(45) 2v6
(46) 3v5 ⟵ 300mV pk-pk playback luminance from pin 47
(47) ⟶ 300mV pk-pk to Luminance delay circuit
(48) 2v8
(49) 1v6 ⟶ Playback & E-E to mixed On screen display AGC circuit
(50) 3v5
(51) 0v Ground
(52) 2v5
(53) 2v2 ⟵ 1v2 pk-pk E-E & Record mode video
(54) ⟵ 5v Supply

TB6515AP

Drum motor driver

Package type B (16 pins)

Pins		Function
(1)	⟶	5v0 pk-pk Drum mixed FG & PG pulses to system control circuit
(2)	⟵	5v Supply
(3) 0v Ground		
(4) 0v Ground		
(5)	⟵	14v Supply
(6) Linked to pin 10		
(7)	⟶	7v0 pk-pk drive to Stator coil 1
(8)	⟶	7v0 pk-pk drive to Stator coil 2
(9)	⟶	7v0 pk-pk drive to Stator coil 3
(10) Linked to pin 6		
(11)	⟵	Drum control Pulse width modulated
(12) 0v Ground		
(13) n/c		
(14) 4v0 pk-pk sawtooth waveform		
(15)	⟵	–ve 400mV pk-pk pulses from Hall effect device
(16)	⟵	+ve 400mV pk-pk pulses from Hall effect device

TBA120T

Sound I.F detector **Package type B (14 pins)**

Pins		Function
(1) 0v Ground		
(2)		
(3) n/c		
(4) n/c		
(5) n/c		
(6)		
(7)		
(8) n/c		
(9) 6MHz Filter		
(10) 6MHz Filter		
(11)	←	12v Supply
(12)	→	1v5 pk-pk Audio output
(13)		
(14)	←	Sound I.F

TC4050BP

6 × invertors/Buffers

Package type B
(16 pins)

Pins		Function
(1)	←	10v Supply
(2)	→	Invertor 1 output
(3)	←	Invertor 1 input
(4)	→	Invertor 2 output
(5)	←	Invertor 2 input
(6)	→	Invertor 3 output
(7)	←	Invertor 3 input
(8) 0v Ground		
(9)	←	Invertor 4 input
(10)	→	Invertor 4 output
(11)	←	Invertor 5 input
(12)	→	Invertor 5 output
(13) n/c		
(14)	←	Invertor 6 input
(15)	→	Invertor 6 output
(16) n/c		

Invertors designated numbers, i.e 1,2, etc are given to distinguish between the individual devices

TD62105

7 × Invertors

Package type B (16 pins)

Pins		Function
(1)	←	Invertor 1 input
(2)	←	Invertor 2 input
(3)	←	Invertor 3 input
(4)	←	Invertor 4 input
(5)	←	Invertor 5 input
(6)	←	Invertor 6 input
(7)	←	Invertor 7 input
(8)	0v Ground	
(9)	n/c	
(10)	→	Invertor 7 output
(11)	→	Invertor 6 output
(12)	→	Invertor 5 output
(13)	→	Invertor 4 output
(14)	→	Invertor 3 output
(15)	→	Invertor 2 output
(16)	→	Invertor 1 output

Invertors designated numbers, i.e 1,2, etc are only given to distinguish between the individual devices

TD62555

4 × invertors

Package type A
(9 pins)

Pins		Function
(1)	←	Invertor 1 input
(2)	←	Invertor 2 input
(3)	←	Invertor 3 input
(4)	←	Invertor 4 input
(5)	0v Ground	
(6)	→	Invertor 4 output
(7)	→	Invertor 3 output
(8)	→	Invertor 2 output
(9)	→	Invertor 1 output

Invertors designated numbers, i.e 1,2, etc are given to distinguish between the individual devices

TDA1543

Digital to analogue convertor

Package type B (8 pins)

Pins		**Function**
(1)	←	Serial clock pulse
(2)	←	Serial data input
(3)	←	Serial data input
(4)	0v Ground	
(5)	←	5v Supply
(6)	2v0 →	Analogue signal output
(7)	Reference voltage	
(8)	2v0 →	Analogue signal output

TDA2451

Vision I.F detector

Package type B
(16 pins)

Pins		Function
(1) 5v8	←	I.F from SAW filter
(2) n/c		
(3) 0v Ground		
(4)	←	Playback mode mute = H
(5)	→	AGC to tuner
(6)	←	AGC control
(7) n/c		
(8) 6v1,Vision detector LC circuit		
(9) 6v1 Vision detector LC circuit		
(10) 2v0		
(11) 3v1	→	2v5 pk-pk Composite video output
(12)		
(13)	←	12v Supply
(14)	←	12v Supply
(15) 0v Ground		
(16) 5v8	←	I.F from SAW filter

TDA4605

Power supply controller **Package type B (8 pins)**

Pins			Function
(1)	0v5	←	Regulation of output control
(2)	1v1	←	Chopper output device current sensing
(3)	1v7	←	Chopper output device voltage sensing
(4)	0v Ground		
(5)	2v7	→	12v pk-pk pulse width modulated output
(6)		←	12v Supply
(7)	1v5	←	Soft start supply
(8)	0v4		

TDA5930

Sound & vision I.F detector

Package type B (16 pins)

Pins		**Function**
(1) 6v2	←	I.F from SAW Filter
(2) 3vo		
(3) 0v Ground		
(4) 2v5		
(5) 3v4		
(6) 2v4	←	RF, AGC control
(7)		
(8) 6v7		
(9) 6v2		
(10) 0v		
(11) 0v	→	Composite video output
(12)	←	12v Supply
(13) 7v0		
(14) 5v6		
(15) 0v Ground		
(16) 6v0	←	I.F from SAW Filter

TDA8501

On screen display generator

Package type B (24 pins)

Pins		**Function**
(1) 2v5	——→	Red chroma
(2)	←——	5v Supply
(3) 2v5	——→	Blue chroma
(4)		
(5) 2v4	——→	Luminance
(6) 2v0		
(7) 2v5	←——	Red chroma
(8)	←——	5v Supply
(9) 2v5	←——	Green chroma
(10) 0v Ground		
(11) 2v5	←——	Blue chroma
(12) 2v2		
(13) 2v5		
(14) 2v0		
(15) 1v0		
(16) 1v8	——→	Composite video output
(17) 0v		
(18) 2v5		
(19) 2v0	——→	Luminance output
(20) 2v5	←——	Luminance from delay line
(21) 2v5		
(22) 2v5	——→	Luminance to delay line
(23) 1v5	←——	Clock pulse
(24) 2v0	←——	Sync' pulses

TDA9715H

Luminance & chroma processor

Package type C (80 pins)

Pins		Function
(1) 2v0	←	30mV pk-pk Record mode Luminance from pin 80
(2) 2v0	←	Luminance white clip adjustment
(3) 3v2		
(4) 1v8		
(5)	←	5v Supply
(6) 4v5	→	1v0 pk-pk FM modulated Luminance output
(7) 1v6	←	Playback luminance level adjustment
(8) 3v0	←	0v5 pk-pk Playback luminance from head circuit
(9) 3v5		
(10) 4v0		
(11)	←	E-E mode = H
(12)	←	Playback LP mode = H, SP = L
(13) 2v6		
(14) n/c		
(15) 2v1		
(16) 2v2	←	Record mode = L, Playback = H
(17) n/c		
(18)	←	5v0 pk-pk flip flop pulse
(19)	←	Trick speed modes = H
(20) 4v5		
(21) 2v5		
(22) 0v Ground		
(23) 2v1		
(24) 0v8		
(25)	←	5v Supply
(26) 2v3	→	4.43MHz Clock pulse to on screen display circuit
(27) 3v1, 4.43MHz xtal voltage controlled oscillator		
(28) 2v9, 4.43MHz xtal voltage controlled oscillator		
(29)	←	5v Supply
(30) 2v6		
(31) 5v0	→	Clock pulse to external Luminance delay circuit
(32) 0v Ground		
(33) 2v2		
(34) 2v4	→	700mV pk-pk record mode chroma to video head circuit
(35) 0v		
(36) 3v5		

(Continued overleaf)

TDA9715H

(37) 0v
(38) 1v7 ◄——— 150mV pk-pk record mode chroma from pin 39
◄——— 100mV pk-pk playback chroma from pin 39
(39) ———► To pin 38
(40) ◄——— 5v Supply
(41) 2v2 ◄——— 300mV pk-pk, 627kHz chroma from video head circuit
(42) 5v0
(43) 0v Ground
(44) 2v2
(45) 2v2
(46) 2v0
(47) 2v0
(48) ———► 700mV pk-pk playback chroma to pin 50 via 2 line delay circuit or pin 49 in MESECAM mode
(49) 2v7 ◄——— From pin 48
(50) 2v7 ◄——— 150mV pk-pk playback chroma from 2 line delay circuit
(51) 0v8
(52) ◄——— MESECAM mode select = H
(53) 2v8 ———► 1v0 pk-pk playback chroma to pin 55 & pin 74 noise cancelling circuit
(54) 2v6
(55) ◄——— 350mV pk-pk playback chroma from pin 53
(56) ◄——— 5v Supply
(57) 2v1 ———► 2v0 pk-pk playback & record modes Luminance output
(58) 0v1 ◄——— Artificial sync' pulses
(59) 2v4
(60) 3v0 ◄——— 1v0 pk-pk record mode composite video input
(61) 2v0
(62) 2v8 ◄——— E-E level adjustment
(63) 0v1
(64) 0v5 ———► 5v0 pk-pk record & playback modes horizontal sync' pulses
(65) 2v8
(66) 0v2
(67) 1v5 ———► 500mV pk-pk playback luminance to pin 68
400mV pk-pk record luminance to pin 68
(68) ◄——— From pin 67
(69) 2v2 ◄——— 500mV pk-pk playback luminance from delay line
400mV pk-pk record luminance from delay line

(Continued opposite)

(70) ——→ 500mV pk-pk playback luminance to delay line
400mV pk-pk record luminance to delay line

(71) 2v0
(72) 2v4
(73) 0v Ground
(74)
(75)
(76) ←—— Picture sharpness control
(77) ——→ 400mV pk-pk record mode Luminance to pin 78
(78) ←—— From pin 77
(79) 2v2
(80) ——→ 1v0 pk-pk record mode luminance to pin 1

TEA2114

Buffered Video signal switch & one 6db gain amplifier — **Package type B (8 pins)**

Pins		Function
(1)	0v Ground	
(2)	⟶	6db gain amplifier output
(3)	⟵	Video 1input
(4)		
(5)	⟵	Video 1 or Video 2 output select
(6)	⟶	Video 1 or Video 2 output
(7)	⟵	12v Supply
(8)	⟵	Video 2 input

TEA5702C

SP Video head amplifier

Package type A (20 pins)

Pins		**Function**
(1) 0v Ground		
(2)	←	Head 2 playback
(3)		
(4)	←	12v Record supply
(5)		
(6)	←	Head 1 playback
(7) 0v Ground		
(8)		
(9)	←	25Hz Head flip flop pulse
(10)		
(11)	←	5v Supply
(12)	←	5v Supply
(13)	→	+ve FM playback output
(14)	→	–ve FM playback
(15)	→	Chroma playback output
(16)	→	Envelope detection
(17) 0v Ground		
(18)	←	Luminance & chroma record
(19)		
(20)	→	Luminance & chroma record to heads

TEA5706

SP & LP Video head amplifier

Package type B (28 pins)

Pins		Function
(1)	←	SP Luminance & chroma record
(2)		
(3)		
(4)	←	SP playback from head 2
(5) 0v Ground		
(6)	←	SP playback from head 1
(7)	←	SP mode playback, head 1 & 2 common connection
(8)	←	LP mode playback, head 1 & 2 common connection
(9)	←	LP playback from head 2
(10) 0v Ground		
(11)	←	LP playback from head 1
(12)		
(13)		
(14)		
(15)		
(16)	←	5v Supply
(17)	→	+ve Luminance FM playback output
(18)	→	–ve Luminance FM playback
(19)	→	Chroma playback
(20)	←	Flip flop pulse
(21)	←	LP / SP switching
(22)		
(23)		
(24)	←	Record mode 12v Supply
(25) 0v Ground		
(26) 0v Ground		
(27)	←	Luminance & chroma record
(28)		

TL8704P

DOC Delay line

Package type B (16 pins)

Pins		**Function**
(1) –ve 5v0, connected to pin 13		
(2) 0v Ground		
(3)	←	5v Supply
(4)		
(5) 0v		
(6) 0v		
(7)	→	Delayed playback FM to noise cancelling & DOC circuit
(8) 4v0		
(9)	←	9v Supply
(10) 0v		
(11) 3v5	←	Playback FM (including drop outs)
(12) 0v		
(13) –ve 5v0, connected to pin 1		
(14) 0v		
(15) 1v8	←	13.9 MHz clock pulse from chroma circuit
(16) 2v3		

U2565

Capstan motor FG amplifier & ACE head driver

Package type B (20 pins)

Pins		Function
(1) 2v5	←	80mV pk-pk from Capstan FG 1
(2) 2v5		
(3) 2v5		
(4) 2v1, connected to control pulse winding ACE head		
(5)		
(6) 2v1	→	1v0 pk-pk playback control pulse
(7) 2v8		
(8) 1v7	←	Playback control pulse from head
(9) 2v0	←	5v0 pk-pk Record control pulse from system micro'
(10) 2v5	←	Control from system micro', positve data pulse = Playback, search, Frame advance, slow forward. negative data pulse = slow reverse
(11) 3v1	→	Playback control pulse phase to system micro'
(12)	←	Record mode = 5v0, Playback mode = 0v
(13) 0v Ground		
(14)	→	Capstan rotation logic to system cont', forward = L, Reverse = H
(15)	→	5v0 pk-pk Capstan to system control / servo
(16)	←	Speed control, X1 = 0v, X4 = 5v0
(17) 2v5		
(18) 2v5		
(19) 2v5	←	80mV pk-pk from Capstan FG 2
(20)	←	5v Supply

U4614B

Chopper power supply control

Package type B
(14 pins)

Pins		**Function**
(1) 1v0	←	Over current sensing threshold
(2) 3v0, Oscillator minimum frequency timing		
(3) 0v	←	8v0 pk-pk Current pulse, on / off control timing
(4) 2v2	→	Timing regulation to pin 8
(5) 3v5 reference		
(6) 0v Ground		
(7) 0v Ground		
(8) 1v6	←	From pin 4
(9) 0v6	→	Low Drive to external chopper transistor
(10) –ve 3v9	→	13v0 pk-pk Drive to external chopper transistor
(11)	←	Start up voltage = 18v0, normal operation = 9v4
(12) 0v	←	Output current sensing
(13) 0v5		
(14) 2v0	→	To pin 1

Pin voltages taken during STOP mode.

UC3842

Chopper power supply controller **Package type B (8 pins)**

Pins		Function
(1)		
(2)	←	Reference voltage
(3)	←	Excessive current sensing
(4)	Oscillator RC timing components	
(5)	0v Ground	
(6)	→	Pulse width modulated drive to chopper transistor
(7)	←	Start up & normal running supply
(8)	Oscillator RC timing components	

UPC1373H

Infra red remote control receiver

Package type A (8 pins)

Pins		Function
(1)	11v7 ⟶	Data output to micro'
(2)	1v4	
(3)	12v, LC peak circuit	
(4)	10,5	
(5)	0v Ground	
(6)	2v0	
(7)	2v6 ⟵	Data from Infra red diode
(8)	⟵	12v Supply

UPC1391H

Sound I.F amplifier & audio detector

Package type A (8 pins)

Pins			Function
(1)		←	12v Supply
(2)	5v7 Sound detector circuit		
(3)	2v5 Sound detector circuit		
(4)	3v0	→	400mV pk-pk Audio
(5)	0v Ground		
(6)		←	I.F Mute control
(7)	3v7	←	6 MHz, 30mV pk-pk Sound carrier
(8)	3v7	←	6 MHz, 30mV pk-pk Sound carrier

UPC1458

Dual operational amplifier

Package type B
(8 pins)

Pins		Function
(1)	⟶	Amplifier 1 output
(2)	⟵	Amplifier 1 inverted input
(3)	⟵	Non inverted input
(4)	0v Ground	
(5)	⟵	Amplifier 2 non inverted input
(6)	⟵	Amplifier 2 inverted input
(7)	⟶	Amplifier 2 output
(8)	⟵	12v Supply

Amplifier designated numbers, i.e 1,2, are only given to distinguish between the individual devices

UPC1473H

Infra red receiver **Package type A (8 pins)**

Pins			Function
(1)	1v0	→	Data output
(2)	2v1		
(3)	4v8		
(4)	3v5		
(5)	0v Ground		
(6)	0v		
(7)	0v7	←	Input from infra red diode
(8)		←	5v Supply

UPC1474H

Infra red receiver

Package type A (9 pins)

Pins		Function
(1)	→	Received data output
(2)	Band pass filter LC circuit	
(3)		
(4)		
(5)	0v Ground	
(6)		
(7)		
(8)	←	Infra red diode input
(9)	←	5v Supply

UPC1490

Remote control receiver amplifier

Package type A (8 pins)

Pins		Function
(1)	←	5v Supply
(2) 5v	→	Data to micro’
(3) 1v4		
(4) 3v5		
(5) 0v Ground		
(6) 1v5		
(7) 2v6		
(8) 2v4	←	Infra red diode receiver

UPC1504C

Drum servo

Package type B
(22 pins)

Pins		**Function**
(1) 2v7	←	3v0 pk-pk FG pulse
(2) 3v1		
(3) 0v Ground		
(4) 5v5		
(5) 1v4	←	Drum motor speed control
(6) 1v5		
(7) 4v5		
(8) 4v0	→	Drum motor speed control
(9) 0v	→	8v0 pk-pk Record mode control pulse to capstan servo
(10)	←	Record mode 9v supply
(11) 0v1	←	1v0 pk-pk vertical pulses
(12)	←	9v5 Supply
(13) 4v8		
(14) 0v Ground		
(15) 3v9	→	Drum phase control
(16) 4v5		
(17)	→	9v0 pk-pk Trapezoid to capstan servo
(18)	←	Record head switching point
(19)	→	Head switching flip flop pulse
(20)	←	Playback head 2 switching point
(21)	←	Playback head 1 switching point
(22) 2v4	←	Drum phase PG pulses

UPC1514

Audio playback & record | **Package type B (18 pins)**

Pins		Function
(1)	←	12v Supply
(2) 5v9	←	Playback audio from ACE head
(3) 5v8		
(4) 5v9	→	Amplified playback audio to pin 5
(5) 5v9	←	From pin 4
(6) 5v9	←	Playback level manual adjustment
(7) 0v Ground		
(8)	←	Audio mute control = H
(9) 5v7	←	Record mode audio input
(10) 5v6		
(11) 5v6	→	Playback audio output & Record mode audio to pin 14
(12) 5v8		
(13)		
(14) 5v6	←	Record mode audio from pin 11
(15) 5v5		
(16) 5v5	→	Record mode audio to head & bias oscillator
(17)	←	Record = H, Playback = L
(18) 0v		

UPC1526C

Capstan Servo

Package type B
(22 pins)

Pins		Function
(1) 3v2	←	5v0 pk-pk FG Pulse
(2) 3v2		
(3) 0v Ground		
(4) 0v Ground		
(5) 1v2		
(6) 5v5		
(7) 1v8		
(8) 4v4		
(9) 4v0	→	Capstan motor speed control
(10) 4v0	→	Capstan motor phase control
(11)	←	9v5 Supply
(12) 0v Ground		
(13) 4v5		
(14)		
(15)	←	9v0 pk-pk Trapezoidal
(16) 0v6		
(17)	←	Record mode 9v Supply
(18) 0v Ground		
(19) 0v4	←	Control pulse originating from pin 20
(20) 3v5	→	6v0 pk-pk Control pulse output
(21) 1v3		
(22) 1v3	←	Control pulse from ACE head

UPC2313

Video Head preamplifier | **Package type B (22 pins)**

Pins		Function
(1)	←	Head switching pulse
(2)	←	Playback mode = 5v0, Record mode = 0v
(3)	2v playback mode, 0v Record mode	
(4)	←	Head 2
(5)	0v Ground	
(6)	←	Head 1
(7)	2v	
(8)	0v Ground	
(9)	5v →	Record current to heads
(10)	5v Record mode	
(11)		
(12)	←	Record mode 9v supply
(13)	0v Playback mode, 1v9 Record mode	
(14)	←	Record mode luminance & chroma FM
(15)	0v playback mode, 5v5 Record mode	
(16)	→	Playback mode chroma output
(17)	0v Ground	
(18)	5v playback, 0v Record mode	
(19)	→	Playback mode luminance output
(20)	→	Playback mode luminance output
(21)	0v6 Playback mode, 0v Record mode	
(22)	←	5v Supply

UPC277G

Dual operational amplifier

Package type B
(8 pins)

Pins		Function
(1)	——→	Amplifier 1 output
(2)	←——	Amplifier 1 inverted input
(3)	←——	Amplifier 1 non inverted input
(4)	0v Ground	
(5)	←——	Amplifier 2 non inverted input
(6)	←——	Amplifier 2 inverted input
(7)	——→	Amplifier 2 output
(8)	←——	5v Supply

UPC324C

4 × Operational amplifiers **Package type B (14 pins)**

Pins		Function
(1)	→	Amplifier 1 output
(2)	←	Amplifier 1 inverted input
(3)	←	Amplifier 1 non inverted input
(4)	←	18v Supply
(5)	←	Amplifier 2 non inverted input
(6)	←	Amplifier 2 inverted input
(7)	→	Amplifier 2 output
(8)	→	Amplifier 3 output
(9)	←	Amplifier 3 inverted input
(10)	←	Amplifier 3 non inverted input
(11) 0v Ground		
(12)	←	Amplifier 4 non inverted input
(13)	←	Amplifier 4 inverted input
(14)	→	Amplifier 4 output

Amplifiers designated numbers, i.e 1,2, etc are given to distinguish between the individual devices

UPC339C

4 × Operational amplifiers

Package type B
(14 pins)

Pins		**Function**
(1)	→	Amplifier 1 output
(2)	→	Amplifier 2 output
(3)	←	12v Supply
(4)	←	Amplifier 2 inverted input
(5)	←	Amplifier 2 non inverted input
(6)	←	Amplifier 1 inverted input
(7)	←	Amplifier 1 non inverted input
(8)	←	Amplifier 4 inverted input
(9)	←	Amplifier 4 non inverted input
(10)	←	Amplifier 3 inverted input
(11)	←	Amplifier 3 non inverted amplifier
(12) 0v Ground		
(13)	→	Amplifier 3 output
(14)	→	Amplifier 4 output

Amplifier designated numbers, i.e 1,2, etc are only given to distinguish between the individual devices

UPC358C

Dual operational amplifier

Package type B (8 pins)

Pins		Function
(1)	→	Amplifier 1 output
(2)	←	Amplifier 1 inverted input
(3)	←	Amplifier 1 non inverted input
(4) 0v Ground		
(5)	←	Amplifier 2 non inverted input
(6)	←	Amplifier 2 inverted input
(7)	→	Amplifier 2 output
(8)	←	12v Supply.

Amplifiers designated numbers, i.e 1,2 are given to distinguish between the individual devices

UPC393C

Dual operational amplifier

Package type B (8 pins)

Pins		Function
(1)	⟶	Amplifier 1 output
(2)	⟵	Amplifier 1 inverted input
(3)	⟵	Amplifier 1 non inverted input
(4)	0v Ground	
(5)	⟵	Amplifier 2 non inverted input
(6)	⟵	Amplifier 2 inverted input
(7)	⟶	Amplifier 2 output
(8)	⟵	12 Supply

Amplifier designated numbers, i.e 1,2, are only given to distinguish between the individual devices

UPC4557

Dual operational audio amplifiers **Package type B (8 pins)**

Pins		Function
(1)	⟶	Amplifier 1 output
(2)	⟵	Amplifier 1 inverted input
(3)	⟵	Amplifier 1 non inverted input
(4)	0v Ground	
(5)	⟵	Amplifier 2 non inverted input
(6)	⟵	Amplifier 2 inverted input
(7)	⟶	Amplifier 2 output
(8)	⟵	12v Supply

Amplifier designated numbers, i.e 1,2, are only given to distinguish between the individual devices

UPC4558

Dual operational audio amplifiers

Package type B (8 pins)

Pins		Function
(1)	⟶	Amplifier 1 output
(2)	⟵	Amplifier 1 inverted input
(3)	⟵	Amplifier 1 non inverted input
(4)	0v Ground	
(5)	⟵	Amplifier 2 non inverted input
(6)	⟵	Amplifier 2 inverted input
(7)	⟶	Amplifier 2 output
(8)	⟵	12v Supply

Amplifier designated numbers, i.e 1,2, are only given to distinguish between the individual devices

UPD4069

6 × Invertors

Package type B
(14 pins)

Pins		Function
(1)	←	Invertor 1 input
(2)	→	Invertor 1 output
(3)	←	Invertor 2 input
(4)	→	Invertor 2 output
(5)	←	Invertor 3 input
(6)	→	Invertor 3 output
(7)	0v Ground	
(8)	→	Invertor 4 output
(9)	←	Invertor 4 input
(10)	→	Invertor 5 output
(11)	←	Invertor 5 input
(12)	→	Invertor 6 output
(13)	←	Invertor 6 input
(14)	←	5v Supply

Invertors designated numbers, i.e 1,2, etc are given to distinguish between the individual devices

UPD4503

6 × 3 state buffers **Package type B (16 pins)**

Pins		Function
(1)	←	Disable pin. (Scan pulses from micro'. Internally connected to buffers 1,2,3, & 4)
(2)	←	Buffer 1 input
(3)	→	Buffer 1 output
(4)	←	Buffer 2 input
(5)	→	Buffer 2 output
(6)	←	Buffer 3 input
(7)	→	Buffer 3 output
(8)	0v Ground	
(9)	→	Buffer 4 output
(10)	←	Buffer 4 input
(11)	→	Buffer 5 output
(12)	←	Buffer 5 input
(13)	→	Buffer 6 output
(14)	←	Buffer 6 input
(15)	←	Disable pin. (Scan pulses from micro'. Internally connected to buffers 5 & 6)
(16)	←	5v Supply

Used in system control switching circuits

Example of basic Buffer operation :
Input pin = H, Disable pin pulse = L., The output pin will be L
Input pin = L, Disable pin pulse = L, the output pin will be L
Input pin = H, Disable pin pulse = H, or the Input pin = H, Disable pin pulse = L, results in the output pin being very high impedance i.e open.

Buffer designated numbers, i.e 1,2, are only given to distinguish between the individual devices

UPD6160C

Capstan & Drum Servo

Package type B (28 pins)

Pins		Function
(1) 2v7	←	12v0 pk-pk Vertical sync' pulse
(2) 0v8	←	Manual tracking adjustment
(3)	←	Serial data
(4)	←	Serial clock pulse
(5) n/c		
(6) 2v5	←	4.43MHz clock pulse
(7) 0v Ground		
(8)	→	SP mode = L
(9) 2v5	→	25Hz head switching pulse
(10) 2v5	→	5v0 pk-pk Drum motor phase control
(11) 2v5	→	5v0 pk-pk Drum motor speed control
(12)	←	Standard play / Long play select
(13) 1v2	←	2v5 pk-pk Drum FG & 5v0 pk-pk Drum PG pulses
(14) 0v Ground		
(15) 0v3	←	Drum PG pulse manual adjustment
(16) 2v7	→	5v0 pk-pk Capstan speed
(17) 2v7	→	5v0 pk-pk Capstan phase
(18)	←	5v0 Supply
(19)	←	From pin 20
(20)	→	1v0 pk-pk Capstan motor FG pulse to pin 19
(21)	←	Capstan motor FG pulse
(22) 2v6		
(23) 2v5		
(24) 2v5	↔	ACE head control pulse
(25) 3v0		
(26) 2v2	→	Amplified playback control pulses to pin 27
(27) 2v2	←	From pin 26
(28) 0v Ground		

UPD6450

On screen character generator

Package type B (18 pins)

Pins		Function
(1) 0v3		
(2) 5v0	←	Clock pulse
(3) 0v1	←	Strobe pulse
(4)	↔	Serial data
(5)	←	5v0 Supply
(6) 2v6		
(7) 2v4, Oscillator LC circuit		
(8) 2v4, Oscillator LC circuit		
(9) 0v Ground		
(10) 5v2, 17.7Mhz xtal		
(11) 5v2, as pin 10		
(12) 4v3	←	Vertical sync'
(13) 4v0	←	Horizontal sync'
(14) 4v0 3v9		
(15) 2v8	←	Character black level control
(16) 5v0	←	Character white level control
(17) 3v4	→	Composite video output
(18) 3v2	←	Composite video input

UPC1484

Tuner band switching

Package type B (18 pins)

Pins			**Function**
(1)		←	Tuning band data
(2)		←	Tuning band data
(3)	7v8		
(4)	5v0	→	Video received signal
(5)	3v8	←	Video input
(6)	0v Ground		
(7)	3v6		
(8)	4v2		
(9)	6v5		
(10)			
(11)			
(12)			
(13)	13v0		
(14)		←	12v Supply
(15)		→	VHF high select
(16)			
(17)		→	VHF low select
(18)		→	UHF select

Drum servo FG frequency to voltage convertor

Package type B (16 pins)

Pins		Function
(1) 2v9		
(2) 2v9	←	Drum FG pulses
(3) 2v9	→	1v0 pk-pk amplified FG pulses to pin 4
(4)	←	From pin 3
(5) 2v8		
(6)	→	FG pulses to internal frequency to voltage convertor pin 15
(7)	←	12v Supply
(8) 0v Ground		
(9)		
(10) 2v9	→	Drum speed drive to motor drive circuit
(11) 1v5		
(12) 2v5		
(13)	←	Drum sampling control
(14) n/c		
(15)	←	4v0 pk-pk Squared Drum FG pulses from pin 6
(16) 2v8		

Capstan servo FG frequency to voltage convertor

(1) 2v9		
(2) 2v9	←	Capstan FG pulses
(3)	→	1v0 pk-pk amplified FG pulses to pin 4
(4)	←	From pin 3
(5) 2v9		
(6)	→	FG pulses to internal frequency to voltage convertor pin 15
(7)	←	12v Supply
(8) 0v Ground		
(9) 5v8		
(10) 2v8	→	Capstan speed drive to motor drive circuit
(11) 1v4		
(12) 2v5		
(13) 4v0	←	Capstan discriminator control
(14) n/c		
(15)	←	4v2 pk-pk squared FG pulse from pin 6
(16) 5v7		

VC2023A

Drum & Capstan servo **Package type B (42 pins)**

Pins		Function
(1)	←	5v Supply
(2)	→	Record mode 4v5 pk-pk control pulse to ACE head
(3) 0v		
(4)	←	Tracking control
(5)	←	Tracking control
(6)		
(7)		
(8) 0v		
(9) 0v		
(10) 0v4	←	Vertical sync' pulse
(11) 1v8	→	4v4 pk-pk Drum flip flop pulse
(12) 1v7		
(13) 0v		
(14) 0v	→	Vertical pulse
(15)	←	Vertical lock control
(16)	←	SP head switching point
(17) 1v2	←	Combined Drum PG & FG pulses
(18) 1v2	←	Combined Drum PG & FG pulses
(19) 2v4	←	Capstan FG pulse
(20)	←	5v0 pk-pk Playback control pulse from ACE head circuit
(21) 0v Ground		
(22) 0v5, xtal system clock		
(23) 0v4, xtal system clock		
(24) 2v1	→	Capstan FG pulse
(25)	←	SP = Lo
(26) 0v Ground		
(27) 2v5		
(28) 0v5	←	Vertical pulse
(29)	←	Record mode = Lo, playback = Hi, Insert = Hi
(30)	←	Record mode = Lo, playback = Hi, insert = Lo
(31)	←	Reverse control
(32)	←	Picture search mode speed control
(33)	←	Picture search mode speed control
(34) 2v0	→	Drum motor speed control to drive circuit
(35) 2v3	→	Drum motor phase control to drive circuit
(36) 0v Ground		
(37) 2v4	→	Capstan phase detection to pin 39
(38) 2v4	→	Capstan speed control to pin 40
(39) 2v4 Capstan phase control from pin 37		
(40) 2v4	←	Capstan speed control from pin 38
(41) 2v4	→	Capstan motor drive to control circuit
(42) 0v Ground		

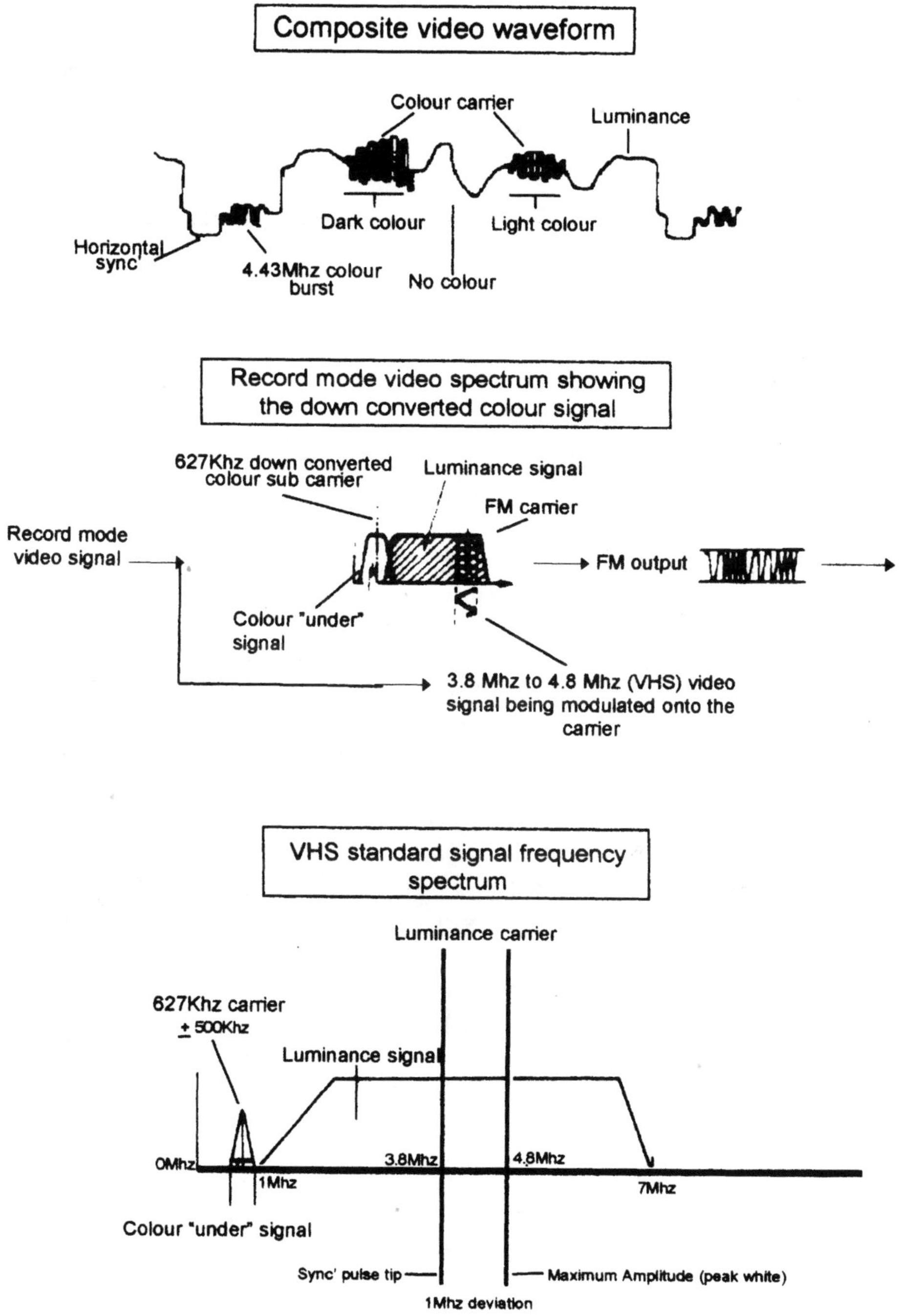
Composite video waveform
Colour carrier
Luminance
Dark colour
Light colour
Horizontal sync'
4.43Mhz colour burst
No colour
Record mode video spectrum showing the down converted colour signal
627Khz down converted colour sub carrier
Luminance signal
FM carrier
Record mode video signal
FM output
Colour "under" signal
3.8 Mhz to 4.8 Mhz (VHS) video signal being modulated onto the carrier
VHS standard signal frequency spectrum
Luminance carrier
627Khz carrier ± 500Khz
Luminance signal
0Mhz
1Mhz
3.8Mhz
4.8Mhz
7Mhz
Colour "under" signal
Sync' pulse tip
Maximum Amplitude (peak white)
1Mhz deviation

Figure 1

Package type A

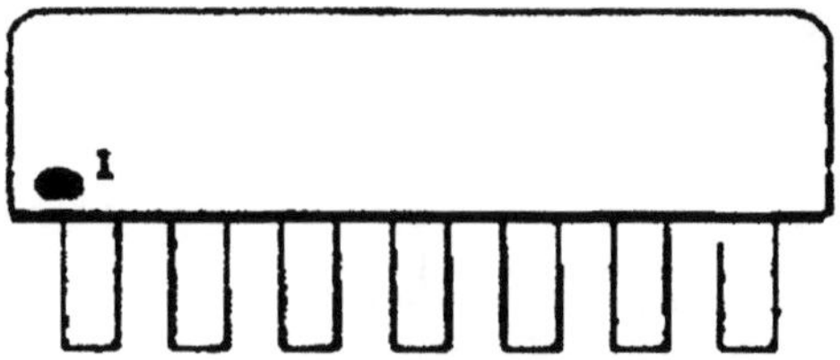

Package type B

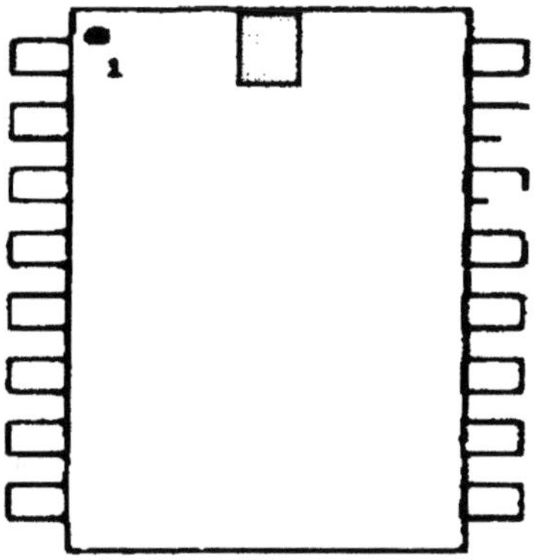

Package type C

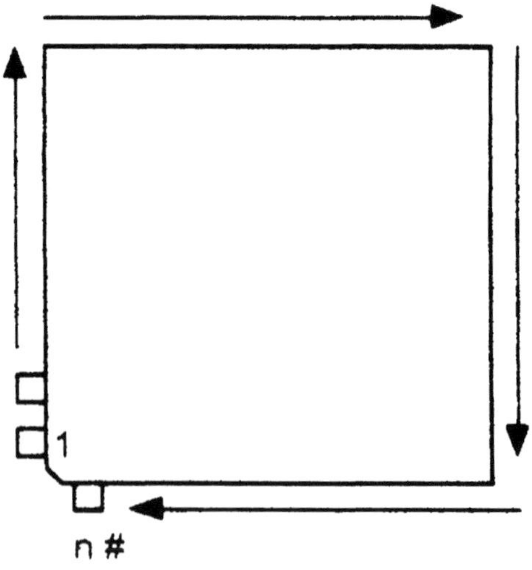

Figure 2 Package types

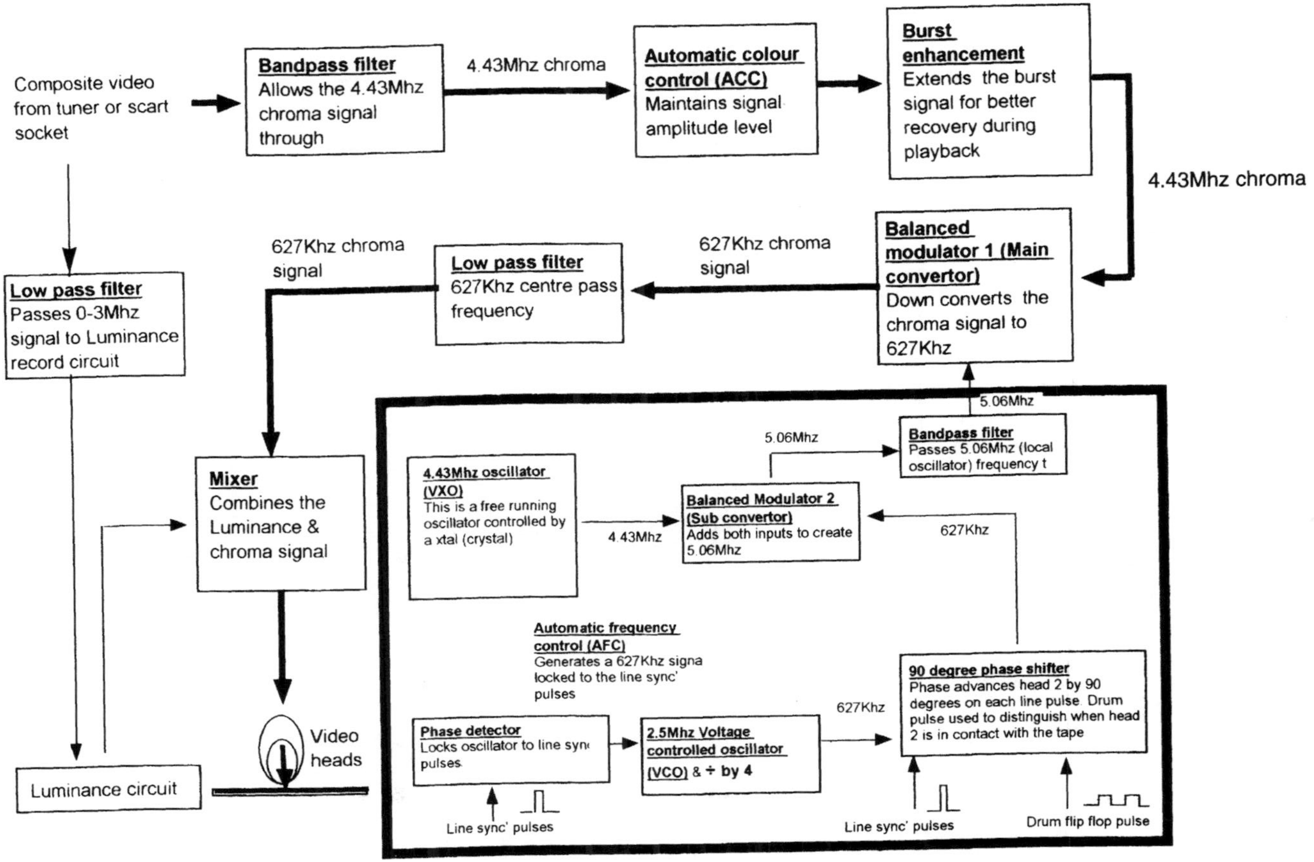

Figure 3 Basic chroma record path

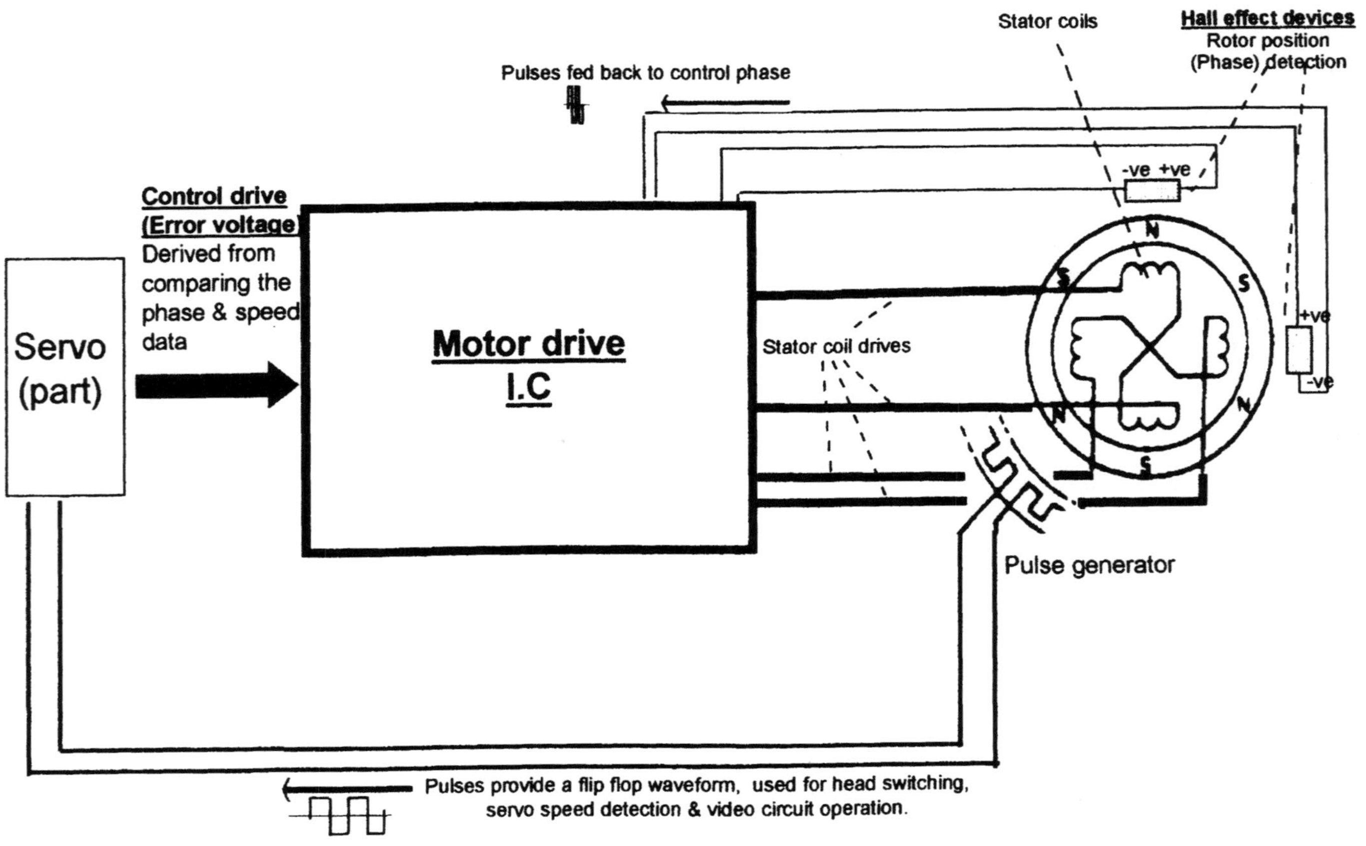

Figure 4 Basic motor drive

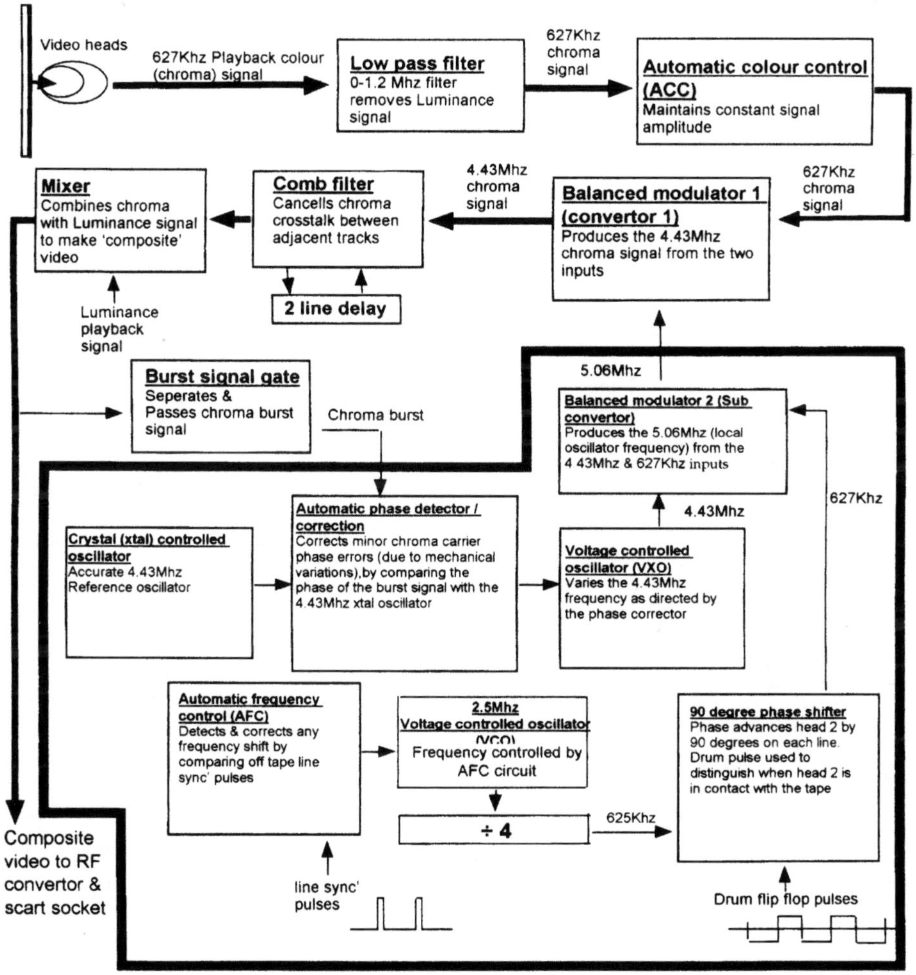

Figure 5 Basic colour (chroma) playback path

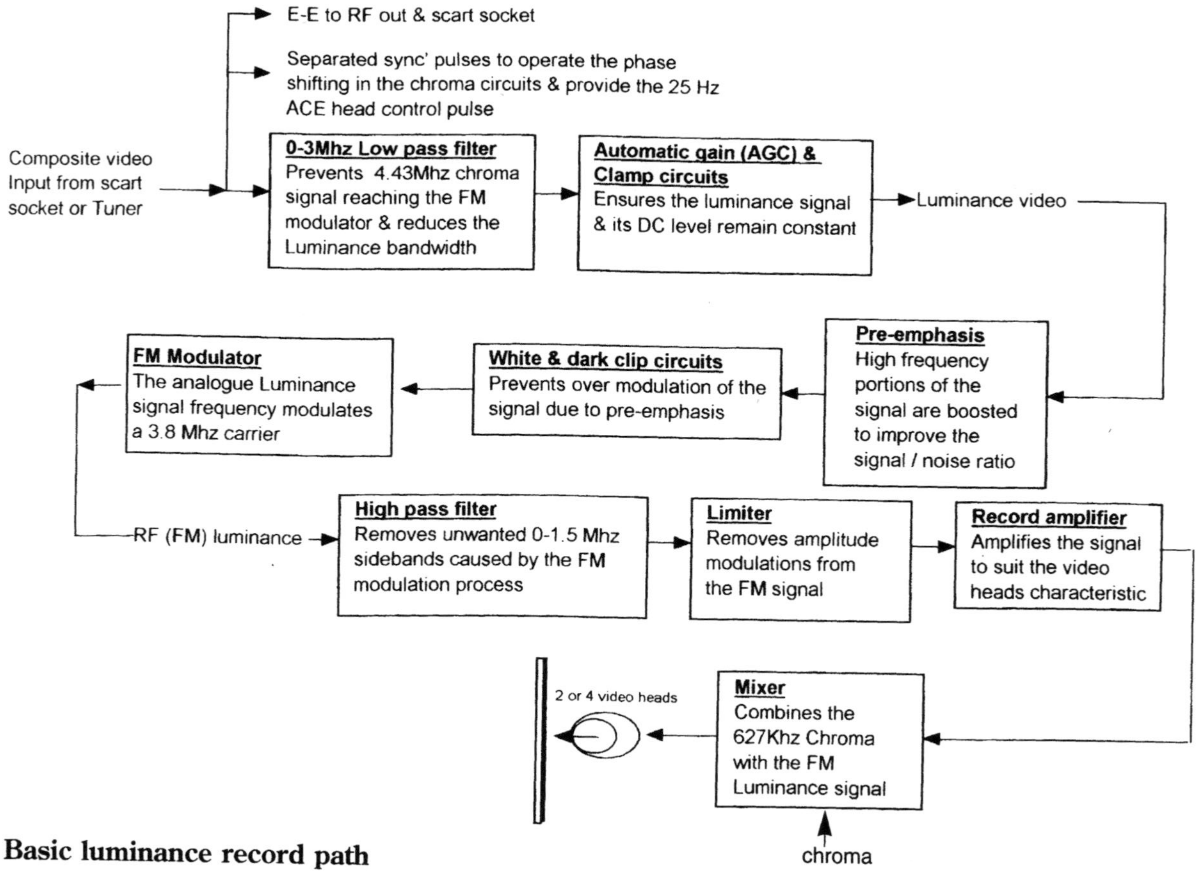

Figure 6 Basic luminance record path

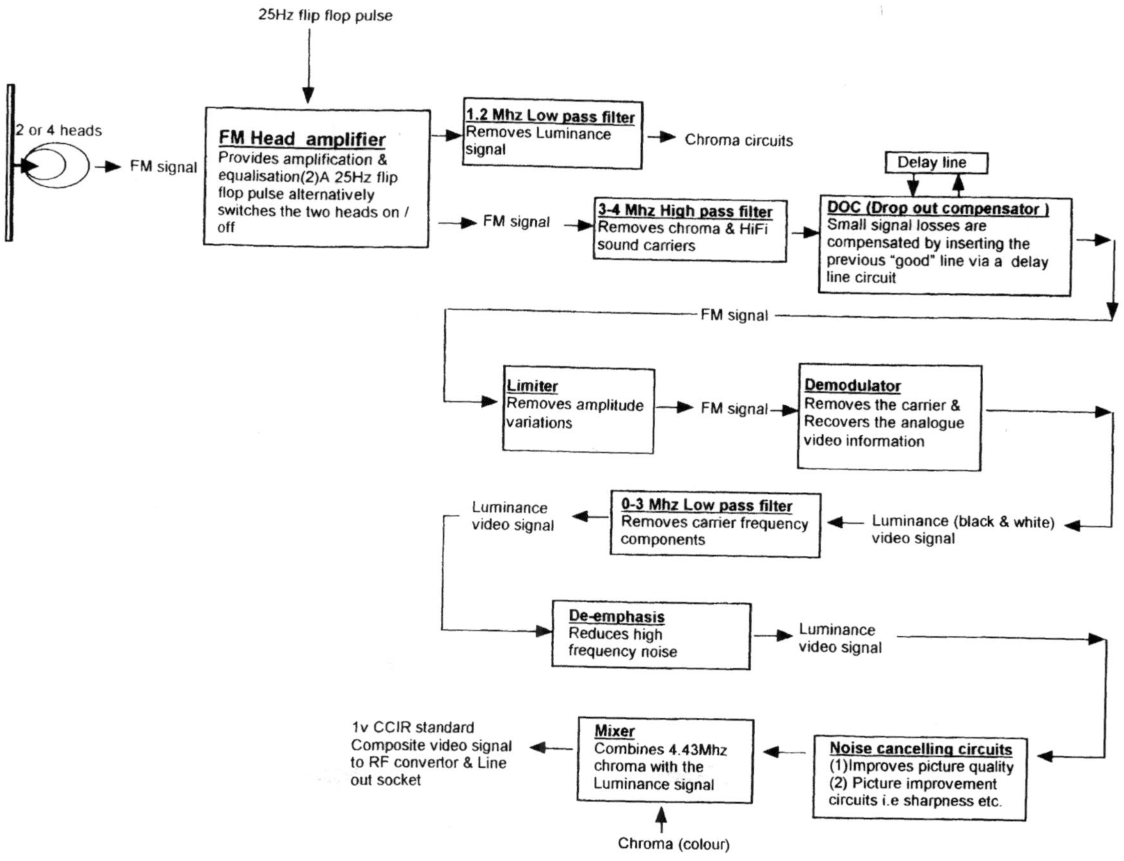

Figure 7 Basic luminance playback path

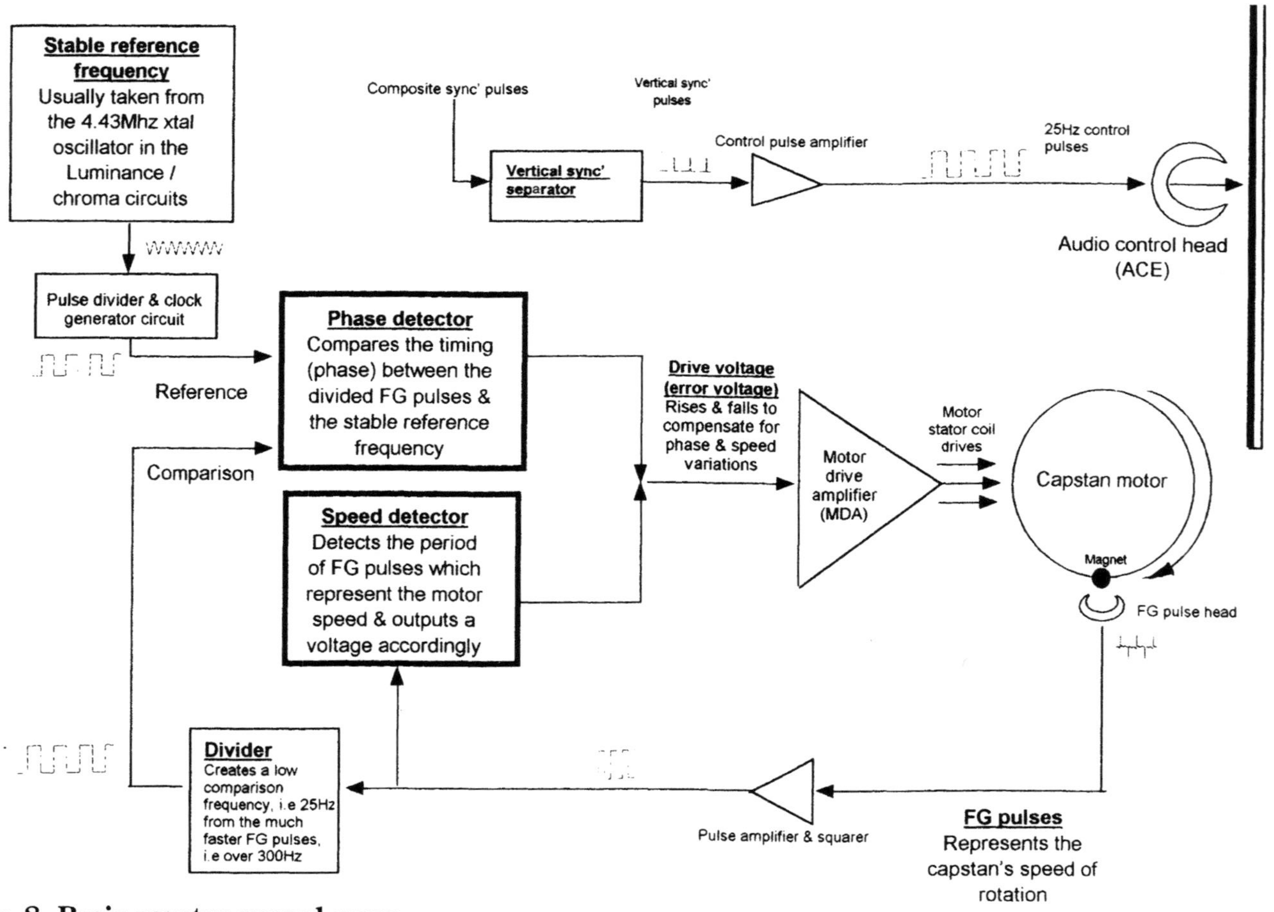

Figure 8 Basic capstan record servo

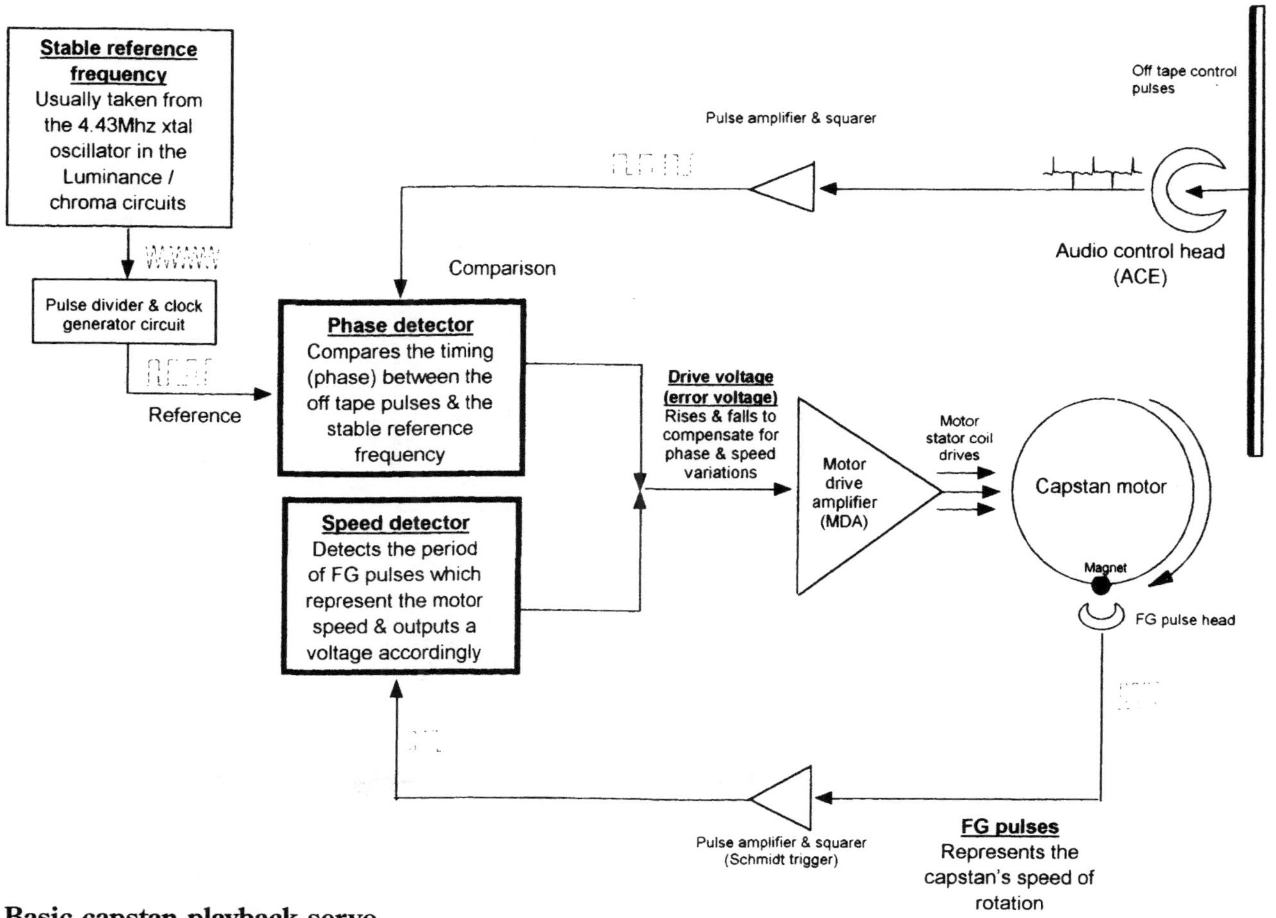

Figure 9 Basic capstan playback servo

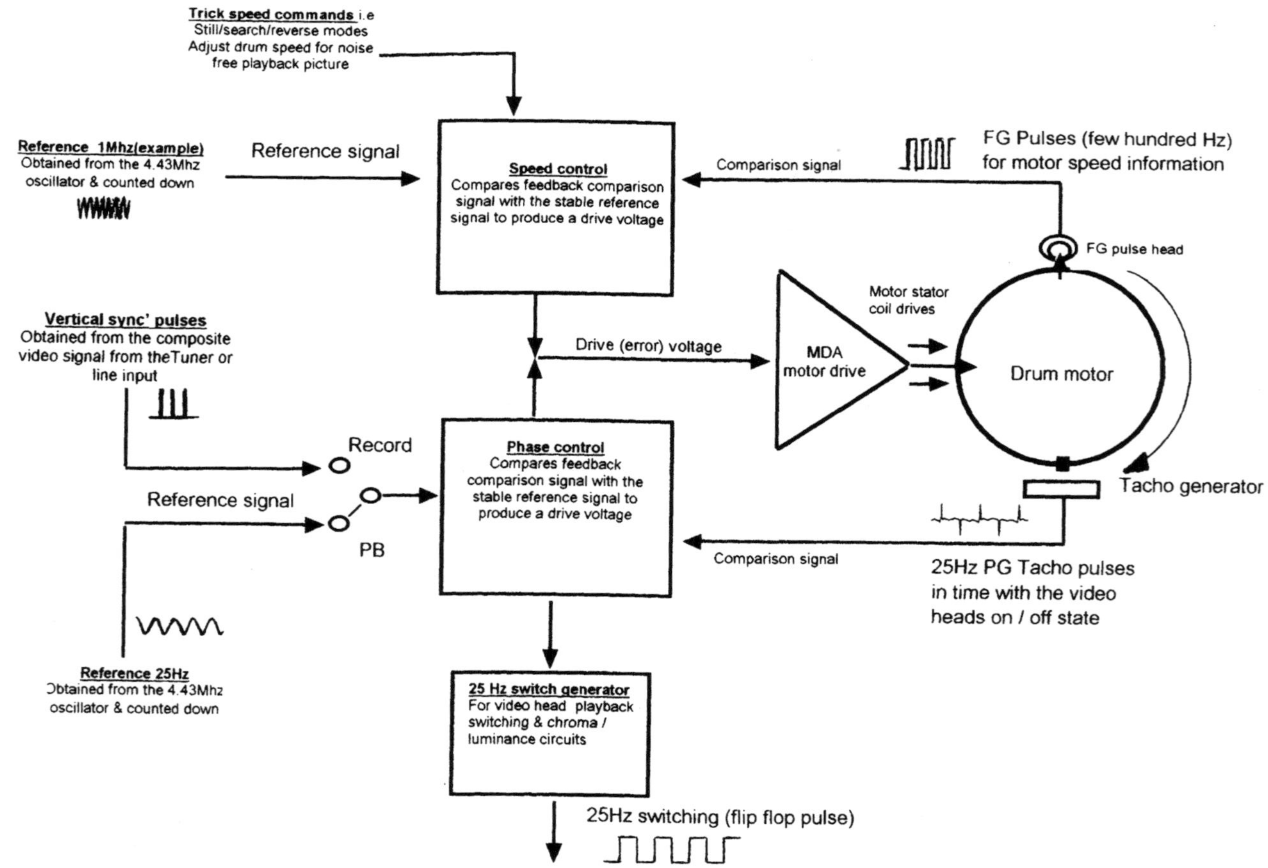

Figure 10 Basic drum (cylinder) motor servo

SYMBOL	TRUTH VALUE TABLE

INVERTER

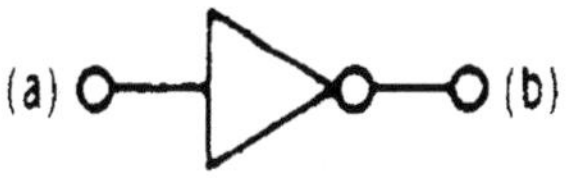

IN	(a)	H	L
OUT	(b)	L	H

COMPARTOR

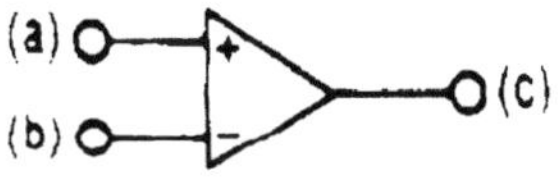

IN	(a) / (b)	(a) > (b)	(a) < (b)
OUT	(c)	H	L

AND CIRCUIT

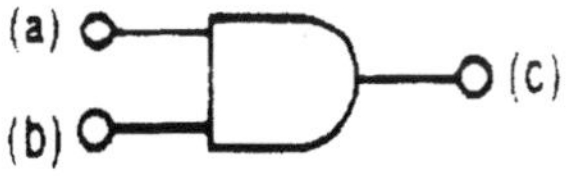

IN	(a)	L	L	H	H
	(b)	L	H	L	H
OUT	(c)	L	L	L	H

OR CIRCUIT

IN	(a)	L	L	H	H
	(b)	L	H	L	H
OUT	(c)	L	H	H	H

THREE STATES BUFFER

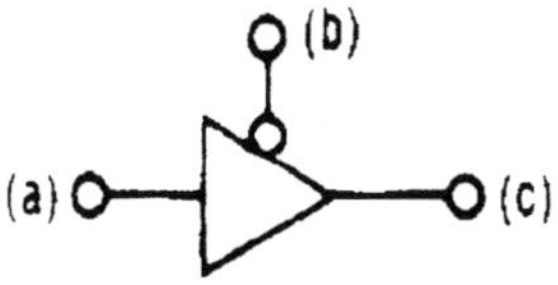

IN	(a)	H	L	H or L
	(b)	L	L	H
OUT	(c)	H	L	※

※ High Impedance

Figure 11 Typical logic devices found in video circuits

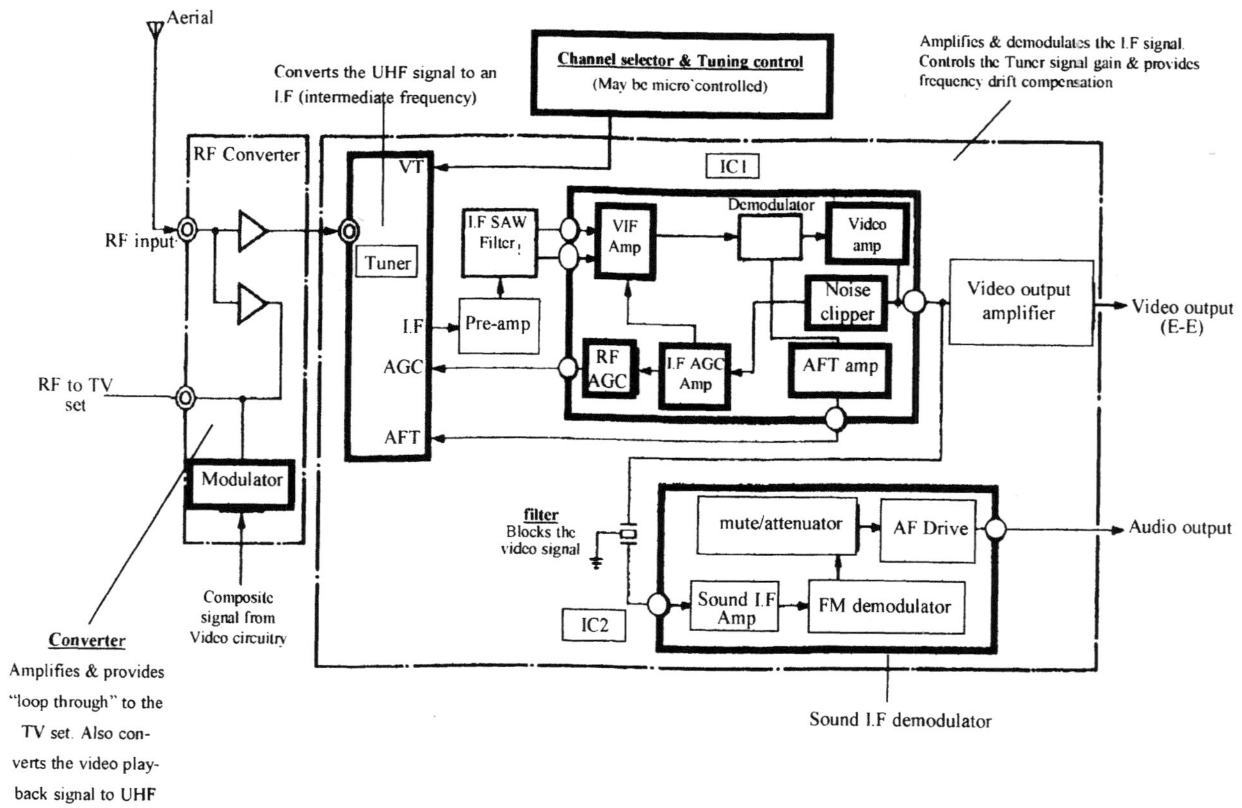

Figure 12 Basic tuner and I.F circuit